DATA LITERACY

DATA LITERACY

HOW TO MAKE YOUR EXPERIMENTS ROBUST AND REPRODUCIBLE

NEIL R. SMALHEISER, MD, PhD

Associate Professor in Psychiatry, Department of Psychiatry and Psychiatric Institute
University of Illinois School of Medicine, USA

ACADEMIC PRESS
An imprint of Elsevier

Academic Press is an imprint of Elsevier
125 London Wall, London EC2Y 5AS, United Kingdom
525 B Street, Suite 1800, San Diego, CA 92101-4495, United States
50 Hampshire Street, 5th Floor, Cambridge, MA 02139, United States
The Boulevard, Langford Lane, Kidlington, Oxford OX5 1GB, United Kingdom

Library of Congress Cataloging-in-Publication Data
A catalog record for this book is available from the Library of Congress

British Library Cataloguing-in-Publication Data
A catalogue record for this book is available from the British Library

ISBN: 978-0-12-811306-6

For information on all Academic Press publications visit our website at
https://www.elsevier.com/books-and-journals

Working together
to grow libraries in
developing countries

www.elsevier.com • www.bookaid.org

Publisher: Mica Haley
Acquisition Editor: Rafael E. Teixeira
Editorial Project Manager: Mariana L. Kuhl
Production Project Manager: Poulouse Joseph
Designer: Alan Studholme

Typeset by TNQ Books and Journals

Cover image credit and illustrations placed between chapters: Stephanie Muscat

Contents

A
DESIGNING YOUR EXPERIMENT

1. Reproducibility and Robustness

2. Choosing a Research Problem

3. Basics of Data and Data Distributions

4. Experimental Design: Measures, Validity, Sampling, Bias, Randomization, Power

5. Experimental Design: Design Strategies and Controls

6. Power Estimation

B
GETTING A "FEEL" FOR YOUR DATA

7. The Data Cleansing and Analysis Pipeline

8. Topics to Consider When Analyzing Data

C
STATISTICS (WITHOUT MUCH MATH!)

9. Null Hypothesis Statistical Testing and the t-Test

10. The "New Statistics" and Bayesian Inference

11. ANOVA

12. Nonparametric Tests

13. Correlation and Other Concepts You Should Know

D
MAKE YOUR DATA GO FARTHER

14. How to Record and Report Your Experiments

15. Data Sharing and Reuse

16. The Revolution in Scientific Publishing

What Is Data Literacy?

Being literate means—literally!—being able to read and write, but it also implies having a certain level of curiosity and acquiring enough background to notice, appreciate, and enjoy the finer points of a piece of writing. A person who has money literacy may not have taken courses in accounting or business, but is likely to know how much they have in the bank, to know whose face is on the 10-dollar bill, and to know roughly how much they spend on the electric bill each month. Many famous musicians have no formal training and cannot read sheet music (Jimi Hendrix and Eric Clapton, to name two), yet they do possess music literacy—able to recognize, produce, and manipulate melodies, harmonies, rhythms, and chord shifts. And data literacy? Almost everyone has some degree of data literacy—one speaks of 1 bird or 2 birds, but never 1.3 birds!

The goal of this book is to learn how a scientist looks at data—how a feeling for data permeates every aspect of a scientific investigation, touching on aspects of experimental design, data analysis, statistics, and data management. After acquiring scientific data literacy, you will not be able to hear about an experiment without automatically asking yourself a series of questions such as: "Is the sampling adequate in size, balanced, and unbiased? What are the positive and negative controls? Are the data properly cleansed and normalized?"

Data literacy makes a difference in daily life too: When a layperson goes to the doctor for a checkup, the nurse tells him or her to take off their shoes and they step on the scale (Fig. 1).

FIGURE 1   A nurse weighs a patient who seems worried — maybe he is thinking about the need for calibration and linearity of the measurement?

When a scientist goes to the doctor's office, before they step on the scale, they tare the scale to make sure it reads zero when no weight is applied. Then, they find a known calibrated weight and put it on the scale, to make sure that it reads accurately (to within a few ounces). They may even take a series of weights that cover the range of their own weight (say, 100, 150, and 200 pounds) to make sure that the readings are linear within the range of its effective operation. They take the weight of their clothes (and contents of their pockets) into account, perhaps by estimation, perhaps by disrobing. Finally, they step on the scale. And then they do that three times and take the average of the three measurements!

This book is based upon a course that I have given to graduate students in neuroscience at the University of Illinois Medical School. Because most of the students are involved in laboratory animal studies, test-tube molecular biological studies, human psychological or neuroimaging studies, or clinical trials, I have chosen examples liberally from this sphere. Some of the examples do unavoidably have jargon, and a basic familiarity with science is assumed. However, the book should be readable and relevant to students and working scientists of any discipline, including physical sciences, biomedical sciences, social sciences, information science, and computer science.

Even though all graduate students have the opportunity to take courses on experimental design and statistics, I have found that the amount of material presented there is overwhelmingly comprehensive. Equally important, the authors of textbooks on those topics come from a different world than the typical student contemplating a career at the laboratory bench. (Hint: There is a hidden but yawning digital divide between the world of those who can program computer code, and those who cannot.) As a result, students tend to learn experimental design and statistics by rote yet do not achieve a basic, intuitive sense of data literacy that they can apply to their everyday scientific life.

Hence this book is not intended to replace traditional courses, texts, and online resources, but rather should be read as a prequel or supplement to them. I will try to illustrate points with examples and anecdotes, sometimes from my own personal experiences—and will offer more personal opinions, advice, and tips than you may be used to seeing in a textbook! On the other hand, I will not include problem sets and will cite only the minimum number of references to scholarly works.

Teaching is harder than it looks.

Acknowledgments

Thanks to John Larson for originally inviting me to teach a course on Data Literacy for students in the Graduate Program in Neuroscience at the University of Illinois Medical School in Chicago. I owe a particular debt of gratitude to the students in the class, whose questions and feedback have shaped the course content over several years. My colleagues Aaron Cohen and Maryann Martone gave helpful comments and corrections on selected chapters. Vetle Torvik and Giovanni Lugli have been particularly longstanding research collaborators of mine, and my adventures in experimental design and data analysis have often involved one or both of them. Finally, I thank my illustrator, Stephanie Muscat, who has a particular talent for capturing scientific processes in visual terms—simply and with humor.

Why This Book?

The scientific literature is increasing exponentially. Each day, about 2000 new articles are added to MEDLINE, a free and public curated database of peer-reviewed biomedical articles (http://www.pubmed.gov). And yet, the scientific community is currently faced with not one, but two major crises that threaten our continued progress.

First, a huge amount of **waste** occurs at every step in the scientific pipeline [1]: Most experiments that are carried out are preliminary ("pilot studies"), descriptive, small scale, incomplete, lack some controls for interpretation, have unclear significance, or simply do not give clear results. Of experiments that do give clear results, most are never published, and the majority of those published are never cited (and may not ever be read!). The original raw data acquired by the experimenter sits in a drawer or on a hard drive, eventually to be lost. Rarely are the data preserved in a form that allows others to view them, much less reuse them in additional research.

Second, a significant minority of published findings cannot be replicated by independent investigators. This is both a crisis of **reproducibility** (failing to find the same results even when trying to duplicate the experimental variables exactly) [2,3] and **robustness** (failing to find similar results when seemingly incidental variables are allowed to vary, e.g., when an experiment originally reported on 6-month-old Wistar rats is repeated on 8-month-old Sprague Dawley rats). The National Institutes of Health and leading journals and pharmaceutical companies have acknowledged the problem and its magnitude and are taking steps to improve the way that experiments are designed and reported [4–6].

What has brought us to this state of affairs? Certainly, a lack of data literacy is a contributing factor, and a major goal of this book is to cover issues that contribute to waste and that limit reproducibility and robustness. However, we also need to face the fact that the culture of science actively encourages scientists to engage in a number of engrained practices that—if we are being charitable—would describe as outdated. The current system rewards scientists for publishing findings that lead to funding, citations, promotions, and awards. Unfortunately, none of these goals are under the direct control of the investigators themselves! Achieving high impact or winning an award is like achieving celebrity in Hollywood: capricious and unpredictable. One would like to believe that readers, reviewers, and funders will recognize and support work that is of high intrinsic quality, but evidence suggests that there is a high degree of randomness in manuscript and grant proposal scores [7,8], which can lead to superstitious behavior [9] and outright cheating. In contrast, **it is within the power of each scientist to make their data solid, reliable, extensive, and definitive in terms of findings**. The interpretation of the data may be tentative and may not be "true" in some abstract or lasting sense, but at least others can build on the data in the future.

In fact, philosophically, there are some advantages to recentering the scientific enterprise around the desire to publish findings that are, first and foremost, robust and reproducible. As we will see, placing a high value on robustness and reproducibility empowers scientists and is part of a larger emerging movement that includes open access for publishing and open sharing of data.

Traditionally, a scientific paper is expected to present a coherent narrative with a strong interpretation and a clear conclusion—that is, it tells a good story and it has a good punch line! The underlying data are often presented in a highly compressed, summarized form, or not presented at all. Recently, however, there has been a move toward considering the raw data themselves to be the primary outcome of a scientific study, to be carefully described and preserved, while the authors' own analyses and interpretation are considered secondary or even dispensable.

We can see why this may be a good idea: For example, let us consider a study hypothesizing that the age of the father (at the time of birth) correlates positively with the risk of their adult offspring developing schizophrenia [10]. Imagine that the raw data consist of a table of human male subjects listing their ages and other attributes, together with a list of their offspring and subsequent psychiatric histories. Different investigators might choose to analyze these raw data in different ways, which might affect or alter their conclusions: For example, one might correlate paternal ages with risk across the entire life cycle, while another might divide the subjects into categorical groups, e.g., "young" fathers (aged 14–21 years), "regular" fathers (aged 21–40 years), and "old" fathers (aged 40+ years). Another

investigator might focus only on truly old fathers, e.g., aged 50 or even 60 years. Furthermore, investigators might correlate ages with overall prevalence of psychiatric illnesses, or any disease having psychotic features, or only those with a stable diagnosis of schizophrenia by the age of 30 years, etc. **Without knowing the nature of the effect in advance, one could defend any of these ways of analyzing the data.**

So, the same data can be sliced and diced in any number of ways, and the resulting publication can look very different depending on how the authors choose to proceed. Even if one accepts that there is some relationship between paternal age and schizophrenia—and this finding has been replicated many times in the past 15 years—it is not at all obvious what this finding "means" in terms of underlying mechanisms. One can imagine that older fathers might bring up their children differently (e.g., perhaps exposing their young offspring to old-fashioned discipline practices). Alternatively, older fathers may have acquired a growing number of point mutations in their sperm DNA over time! Subsequent follow-up studies may attempt to characterize the relationship of age to risk in more detail, and to test hypotheses regarding which possible mechanisms seem most likely. And of course, the true mechanism(s) might reflect genetic or environmental influences that are not even appreciated or known at the time that the relation of age to risk was first noticed.

To summarize, the emerging view is that the bedrock of a scientific paper is its **data**. The authors' presentation and analysis of the data, resulting in its primary **finding**, is traditionally considered by most scientists to be the outcome of the paper, and it is this primary finding that ought to be robust and reproducible. However, as we have seen, the primary finding is a bit more subjective and removed from the data themselves, and according to the emerging view, it is NOT the bedrock of the paper. Rather, it is important that independent investigators should be able to view the raw data to reanalyze them, or compare or pool with other data obtained from other sources. Finally, the authors' **interpretation** of the finding, and their general conclusions, may be insightful and point the way forward, but should be taken with a big grain of salt.

The status quo of scientific practice is changing, radically and rapidly, and it is important to understand these trends to do science in the 21st century. This book will provide a roadmap for students wishing to navigate each step in the pipeline, from hypothesis to publication, during this time of transition. Do not worry, this roadmap won't turn you into a mere data collector. Finding novel, original, and dramatic findings, and achieving breakthroughs will remain as important as ever.

References

[1] Chalmers I, Glasziou P. Avoidable waste in the production and reporting of research evidence. *Lancet* July 4, 2009;**374**(9683):86–9. http://dx.doi.org/10.1016/S0140-6736(09)60329-9.

[2] Ioannidis JP. Why most published research findings are false. *PLoS Med* August 2005;**2**(8):e124.

[3] Leek JT, Jager LR. Is most published research really false? *bioRXiv* April 27, 2016. http://dx.doi.org/10.1101/050575.

[4] Landis SC, Amara SG, Asadullah K, Austin CP, Blumenstein R, Bradley EW, Crystal RG, Darnell RB, Ferrante RJ, Fillit H, Finkelstein R, Fisher M, Gendelman HE, Golub RM, Goudreau JL, Gross RA, Gubitz AK, Hesterlee SE, Howells DW, Huguenard J, Kelner K, Koroshetz W, Krainc D, Lazic SE, Levine MS, Macleod MR, McCall JM, Moxley 3rd RT, Narasimhan K, Noble LJ, Perrin S, Porter JD, Steward O, Unger E, Utz U, Silberberg SD. A call for transparent reporting to optimize the predictive value of preclinical research. *Nature* October 11, 2012;**490**(7419):187–91. http://dx.doi.org/10.1038/nature11556.

[5] Hodes RJ, Insel TR, Landis SC. On behalf of the NIH blueprint for neuroscience research. The NIH toolbox: setting a standard for biomedical research. *Neurology* 2013;**80**(11 Suppl. 3):S1. http://dx.doi.org/10.1212/WNL.0b013e3182872e90.

[6] Begley CG, Ellis LM. Drug development: raise standards for preclinical cancer research. *Nature* March 28, 2012;**483**(7391):531–3. http://dx.doi.org/10.1038/483531a.

[7] Cole S, Simon GA. Chance and consensus in peer review. *Science* November 20, 1981;**214**(4523):881–6.

[8] Snell RR. Menage a quoi? Optimal number of peer reviewers. *PLoS One* April 1, 2015;**10**(4):e0120838. http://dx.doi.org/10.1371/journal.pone.0120838.

[9] Skinner BF. Superstition in the pigeon. *J Exp Psychol* April 1948;**38**(2):168–72.

[10] Brown AS, Schaefer CA, Wyatt RJ, Begg MD, Goetz R, Bresnahan MA, Harkavy-Friedman J, Gorman JM, Malaspina D, Susser ES. Paternal age and risk of schizophrenia in adult offspring. *Am J Psychiatry* September 2002;**159**(9):1528–33.

How many potential new discoveries are filed away somewhere, unpublished, unfunded, and unknown?

DESIGNING YOUR EXPERIMENT

1

Reproducibility and Robustness

BASIC TERMS AND CONCEPTS

An experiment is said to be successfully **replicated** when an independent investigator can repeat the experiment as closely as possible and obtain the same or similar results. Let us see why it is so surprisingly difficult to replicate a simple experiment even when no fraud or negligence is involved. Consider an (imaginary) article that reports that Stanford college students majoring in economics are less likely to have signed an organ donor card than students majoring in social work. The authors suggest that students in the "caring" professions may be more altruistic than those in the "money" professions. What does it mean to say that this article is **reproducible**? Following the major sections of a published study (see Box 1.1), one must separate the question into five parts:

What does it mean to replicate the **data** obtained by the investigators?
What does it mean to replicate the **methods** employed by the investigators?
What does it mean to replicate the **findings**?
What does it mean to say that the findings are **robust** or **generalizable**?
What does it mean to replicate the interpretation of the data, i.e., the authors' **conclusion**?

Replicating the Data

We will presume that the investigators took an adequately large sample of students at Stanford—that they either (1) examined all students (or a large unbiased random sample), and then restricted their analysis to economics majors versus social work majors, or (2) sampled only from these two majors. We will presume that they discerned whether the students had signed an organ donor card by asking the students to fill out a self-report questionnaire. **Reproducibility** of the data means that if someone took another random sample of Stanford students and examined the same majors using the same methods, **the distribution of the data would be (roughly) the same**, that is, there would be no statistically significant differences between the data distributions in the two data sets. In particular, the demographic and baseline characteristics of the students should not be essentially different in the two data sets—it would be troubling if the first experiment had interviewed 50% females and the

---

BOX 1.1

THE NUTS AND BOLTS OF A SCIENTIFIC REPORT

A typical article will introduce a **problem**, which may have previously been tackled by the existing literature or by making a new **observation**. The authors may pose a **hypothesis** and outline an experimental plan, either designed to test the hypothesis conclusively or more often to shed more light on the problem and constrain possible explanations. After acquiring and analyzing their **data**, the authors present their **findings** or **results**, discuss the implications and limitations of the study, and point out directions for further research.

As we will discuss in later chapters in detail, the data associated with a study is not a single entity. The **raw data** acquired in a study represent the most basic, unfiltered data, consisting of images, machine outputs, tape recordings, hard copies of questionnaires, etc. This is generally transcribed to give numerical summary measurements and/or textual descriptors (e.g., marking subjects as male vs. female). Often each sample is assigned one row of a spreadsheet, and each measure or descriptor is placed in a different column. This spreadsheet is still generally referred to as **raw data**, even though the original images, machine reads, or questionnaire answers have been transformed, and some information has been filtered and possibly lost.

Next, the raw data undergo successive stages of **data cleansing**: Some experimental runs may be discarded entirely as unreliable (e.g., if the control experiments in these runs did not behave as expected). Some data points may be missing or suspicious (e.g., suggestive of typographical errors in transcribing or faulty instrumentation) or anomalous (i.e., very different from most of the other points in the study). How investigators deal with these issues is critically important and may affect their overall results and conclusions, yet different investigators may make very different choices about how to proceed. Once the data points are individually cleansed, the data are often **thresholded** (i.e., points whose values are very low may be excluded) and **normalized** (e.g., instead of considering the raw magnitude of a measurement, the data points may be **ranked** from highest to lowest and the ranks used instead) and possibly data points may be grouped into **bins** for further analysis. Again, this can be done in many different ways, and the choice of how to proceed may alter the findings and conclusions. It is important to preserve ALL the data of a study, including each stage of data transformation and cleansing, to allow others to replicate, reuse, and extend the study.

The findings of a study often take the form of comparing two or more experimental groups with regard to some measurement or parameter. Again, the findings are not a single entity! At the first level, each experimental group is associated with that measurement or parameter, which is generally summarized by the sample size, a mean (or median) value, and some indication of its variability (e.g., the standard deviation, standard error of the mean, or confidence intervals). These represent the most basic findings and should be presented in detail.

Next, two or more experimental groups are often compared by measuring the absolute **difference** in their means or the **ratio** or **fold difference** of the two means. (Although

---

BOX 1.1 *(cont'd)*

---

the difference and the ratio are closely related, they do not always convey the same information—for example, two mean values that are very small, say 0.001 vs. 0.0001, may actually be indistinguishable within the margin of experimental error, and yet their ratio is 10:1, which might appear to be a large effect.) Ideally, both the differences and the ratios should be analyzed and presented.

Finally, especially if the two experimental groups appear to be different, some statistical test(s) are performed to estimate the level of statistical significance. Often a P-value or F-score is presented. Statistical significance is indeed an important aspect, but to properly assess and interpret a study, a paper should report ALL findings—the sample size, mean values and variability of each group, and the absolute differences and fold differences of two groups. Only then should the statistical significance be presented.

This brief outline shows that the "data" and "findings" of even the simplest study are surprisingly complex and include a mix of objective measurements and subjective decisions. The current state of the art of publishing is such that rarely does an article preserve all of the elements of the data and findings transparently, which makes it difficult, if not impossible, for an outside laboratory to replicate a study exactly or to employ and reuse the data fully for their own research. It is even common in certain fields to present ONLY the P-values as if those are the primary findings, without showing the actual means or even fold differences! Clearly, as we proceed, we will be advising the reader on proper behavior, regardless of whether this represents current scientific practice!

---

replication experiment only 20% females or if the first experiment interviewed a much higher proportion of honors students among the economics majors versus the social work majors, and this proportion was reversed in the replication experiment.

Replicating the Methods

This refers to detailed, transparent **reporting** and **sharing** of the methods, software, reagents, equipment, and other tools used in the experiment. We will discuss reporting guidelines in Chapter 14. Here it is worth noting that many reagents used in experiments cannot be shared and utilized by others, because they were generated in limited amounts, are unstable during long-term storage, are subject to proprietary trade secrets, etc. Freezers fail and often only then does the experimenter find out that the backup systems thought to be in place (backup power, backup CO_2 tanks) were not properly installed or maintained. Not uncommonly, reagents (and experimental samples) become misplaced, uncertainly labeled, or thrown out when the experimenter graduates or changes jobs.

Replicating the Findings

The stated finding is that students majoring in economics are less likely to have signed an organ donor card than students majoring in social work. That is, there is a **statistically significant difference** between the proportion of economics majors who have signed cards versus the proportion of social work majors who have signed cards. But note that a statement of statistical significance is actually a derived parameter (the *difference* between two primary effects or experimental outcomes), and it is important to state not only the fact that a difference exists but also to report each measured effect size(s), which more closely reflect the actual experimental data:

Major	N	Signed Card	Proportion	95% Confidence Interval
Economics	105	22	21%	15%–27%
Social work	101	43	43%	36%–50%

In particular, to say that the reported finding can be replicated, the replication experiment should find similar **effect sizes**, i.e., signed cards in "close to" 21% of economics majors and "close to" 43% of social work majors. One can imagine carrying out a second experiment and finding that 55% of economics majors have signed organ donor cards versus 90% of social work majors. The effect sizes in the latter case are entirely different, and so one cannot say that the results are similar, even though there is still a huge difference between groups that we presume will be statistically significant.

Effect sizes and statistical significance are very different aspects of a study! It is a matter for careful judgment to decide whether, in a given experiment, replicating a statistically significant difference between groups is more relevant and important than replicating the effect sizes per se. As we will see later, a failure to reproduce the same baseline is an important red flag that there may be problems or confounds in the study. However, sometimes a fluctuating baseline is to be expected. For example, the US economy regularly undergoes cycles of growth and recession, so any random sampling of Americans is likely to show a significant difference in the percentage of unemployed individuals at different times. This needs to be taken into account, although it may or may not affect the results or interpretation in the context of a particular study.

Robustness and Generalizability

What does it mean to say that the organ donor card finding is robust or that the finding can be generalized? **Robustness** and **generalizability** are closely related concepts: A finding is robust if it is observed regardless of small or incidental changes in experimental variables, whereas a finding is generalizable if it holds across a wider variety of situations.

For example, the original study may have collected data on a Monday in April. One can well imagine that subjects might be more cooperative or candid depending on the timing relative to exams or holidays or even the zeitgeist (e.g., just prior to vs. just after the 9/11 attacks). As well, their cooperation may vary according to the appearance or training of the interviewer (Fig. 1.1). If the data or findings are altered significantly depending on such variables, one would say that the study is not robust. Similarly, if the results vary significantly depending on the method of ascertainment (self-report questionnaire vs. asking the students to show their organ donor cards vs. cross-checking their names against the state organ donor list), the study might be reproducible but is not robust. And finally, one would need to examine a variety of "money" versus "caring" professional majors (e.g., accounting vs. nursing) before one would be entitled to assess the generalizability of the findings.

FIGURE 1.1 Interviewees may cooperate or respond differently depending on the appearance or training of the interviewer.

A. DESIGNING YOUR EXPERIMENT

Replicating the Conclusion

Missing so far from this discussion is the actual currency of science—that is, the questions, hypotheses, and tentative models that existed in the minds of the investigators and motivated them to study economics versus social work majors in the first place. Their conclusion indicates that they are using a person's major, which can be objectively measured, as a **proxy** for something else that cannot be directly measured, but which plays an explanatory role in their models. That is, they suspect that choice of profession may correlate with an individual's "selfish" versus "caring" personality profile and may influence their likelihood of exhibiting altruistic behavior.

People who pick different majors are likely to differ in many ways, some of which may directly or indirectly reflect their personalities. For example, economics and social work majors at the same school might differ overall in gender mix, age distribution, religious affiliation, socioeconomic class, ethnicity, immigrant status, political leanings, perhaps even sexual orientation. If the purpose of the study is to correlate organ donor card participation with the person's major per se, then one should examine the correlation of card participation with major after taking into account (i.e., statistically removing the contributions from) all these other factors. But if the purpose is to use a person's major as a proxy for their personality, then it would not be appropriate to remove the effects of all of the ancillary factors that also contribute to or interact with a person's personality. Instead, it would be better to combine these factors together into a single composite profile that should be a better proxy for personality than a person's major by itself. It is no simple task to say what the right way to design an experiment is.

These ancillary factors are also important to take into account when carrying out replication studies. For example, the student profiles of economics majors at Stanford, a private, elite school on the West Coast, are likely to differ in many ways from economics majors at University of Illinois at Chicago (UIC), a large, public, urban university in the Midwest largely attended by children of immigrants. (And in real life, Stanford does not actually offer a major in social work at all.) It is not at all obvious whether economics majors have the same personality patterns at Stanford versus UIC. So if we find that the same experiment carried out at Stanford gives different results at UIC, we cannot say that the authors' conclusions are necessarily wrong. That is, it is still possible that students in the "caring" professions may be more altruistic than those in the "money" professions. We can definitely say that the original experiment is not robust, however, and this would cast doubt on the utility of using a person's major as a proxy for their personality profile or as a correlate of their behavior.

REPRODUCIBILITY AND ROBUSTNESS TALES TO GIVE YOU NIGHTMARES

The traditional purpose of a scientific article is to describe novel and believable observations or experiments in the form of "We did X and observed Y," including appropriate controls to demonstrate that their results are not the result of gross error or confounds. The traditional emphasis has been on showing that the results have **validity**, i.e., are "correct" or "true" descriptions of what happened, within the margins of error or limitations of the assays involved. Yet validity is quite different from reproducibility or robustness.

A. DESIGNING YOUR EXPERIMENT

For example, let us consider a classic article in the neuroscience literature. An epileptic patient, H.M., underwent removal of the medial temporal regions of the brain on both sides to stop his seizures and was observed to have a severe loss of his ability to make new memories [1]. To say that this finding is **valid** means asking whether the authors chose appropriate psychological tests, performed them properly, and had ruled out alternative explanations. (Was he pretending? Was he simply unmotivated or uncooperative?) To say that the finding is **reproducible** means that others examining H.M. independently would come to the same conclusion. (Indeed, his memory deficit lasted for his entire lifetime, and he has been the subject of many publications over the years [2].) To say the finding is **robust** means examining multiple patients given the same surgery and asking whether the same memory deficits are observed consistently. In their initial report [1], Scoville and Milner (1957) looked at 10 patients and a common pattern emerged: Those who had removal of the hippocampus bilaterally showed severe memory deficits, whereas those whose surgery affected other medial temporal structures did not. Historically, this provided important evidence linking the hippocampus to the acquisition of new memories.

You might think that an article that has validity, which accurately reports an experimental finding, would also tend to exhibit reproducibility and robustness. Yet as the following paragraphs will reveal, there exist entire extensive and popular fields of biomedical science that are filled with articles reporting findings that are novel and valid—they satisfy peer reviewers and appear in high-quality journals—and still, as a rule, they neglect to address basic criteria of reproducibility and robustness. This severely limits the value of each article. For example, imagine if Scoville and Milner had published 10 different articles, at 10 different times, and in 10 different journals, examining one patient each! Discerning a common pattern would have been rendered much more difficult.

Cell Culture Experiments

One of the most popular experimental approaches in biology involves culturing cells and characterizing their response when specific conditions are altered (e.g., when they are exposed to drugs, signaling proteins, nutrients, surfaces, or gene transfer). The cells may be freshly extracted from tissues, or more often, they are propagated as continuously growing cell lines (which, in turn, may be derived either from normal or cancerous cells). Cell culture experiments exemplify well the daunting challenges that scientists face in attempting to make their experiments not only valid (an accurate description of how the cells responded in their hands) and replicable (reproducible by others when the experiment is exactly duplicated) but also robust and generalizable, that is, applicable beyond one setting and contributing to knowledge in some larger or more general sense.

The first problem is knowing what kind of cells one is studying! For example, during the 1960s and 1970s, it became apparent that one-third or more of the cell lines thought to represent normal organ-derived cell types were in fact contaminated with (or totally taken over by) the fast-growing human cancer, HeLa cell line [3,4] (Fig. 1.2). Thousands of articles purporting to report experiments in one type of cell (e.g., embryonic lung fibroblasts or kidney epithelial cells) were in fact performed in HeLa cells! To my knowledge, few if any of these articles have been retracted or corrected. The fact that there are many commercial services that provide cell line authentication only underscores the fact that the problem persists to

FIGURE 1.2   **Photomicrograph of cultured HeLa cells** stained with DAPI (green) and phalloidin (red). DAPI stains the nucleus of the cells, whereas phalloidin visualizes actin filaments. The image depicts the position of nucleus and architecture of actin filaments. *Modified from Arnatbio, https://commons. wikimedia.org/wiki/File:Immunostained_HELA_cells. tif#/media/File:Immunostained_HELA_cells.tif.*

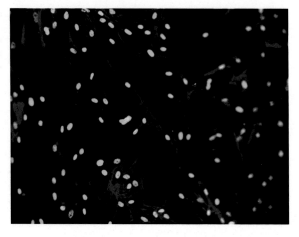

this day. Investigators still do not routinely provide evidence that they have authenticated their cells at the time the studies were done, nor do journals or reviewers insist upon such assurance. Even HeLa cells are not a single entity. Different stocks of HeLa cells growing in different laboratories diverge substantially from each other in terms of their DNA content and their patterns of gene expression and chromosomal instability, which would be expected to result in laboratory-specific responses to experimental stimuli [5].

The same cell line may also show totally different characteristics in the same laboratory depending on how it is cultured. For example, I remember studying cells that normally grow as a sheet upon tissue culture plastic dishes, and encountering a particular batch of tissue culture plastic which caused them to grow in small scattered clumps instead. I reported this to the company, and 2 years later I was surprised to find another lot of plastic that produced the same strange behavior. It turns out that they had shipped me the same defective lot again. When I asked why they had not pulled that lot from their inventory, they replied that no one else had complained! Not all investigators examine the morphology of their cultured cells routinely, especially when they are only concerned with biochemical endpoints. Who knows how many published experiments employed that batch of plastic dishes and obtained results that were specific to that lot number?

These problems might not be so bad if cells tend to respond to the same stimuli, more or less in the same manner. Au contraire—if one examines, say, five different human lung cancer lines and examines their response to some treatment (e.g., the cytokine protein Interleukin-6), one will find major quantitative and even qualitative differences in how they respond, the effective dose—response range, the time course of response, and even whether they respond at all. Conversely, the responses of the exact same cell type to the same treatment can vary widely according to the ingredients of its growth media (which include a mix of proteins, ions, buffers, etc.) as well as the temperature, oxygen, and carbon dioxide concentrations of the culture chamber. Generally the cells are seeded at low density and allowed to grow toward confluence (i.e., filling the entire surface of a culture dish). The physiology of cells varies in a major way according to their density, the state of confluence, the position in the cell cycle, and the time since last feeding, to name a few parameters. They even vary according to the details of how dishes are fed with new media—Is the existing media removed entirely or is only half removed and an additional half of new media added? Even the nature of the vehicle used to dissolve the drug (DMSO vs. ethanol vs. saline) matters.

If the cells are being extracted, say, to measure levels of a particular cellular protein, it is critical to consider how the cells are harvested. Results may differ depending on whether the cells are detached using trypsin versus EDTA versus mechanically scraped versus pipetted. Are the cells spun down to collect them, and if so, at what speed for how long? If cells are extracted for protein, is a detergent used? If so, which one, at what concentration, treated for how long at what temperature; and are sodium, calcium, calcium chelators, or magnesium ions present? Each of these parameters can make a major difference: The amount of sodium (as a determinant of ionic strength) affects the ease of detaching proteins from each other that rely on ionic bonds; calcium is a cofactor in cell—cell binding through its control of integrin and cadherin receptors and can affect stability of protein—protein complexes such as polyribosomes; and magnesium is necessary for stability of RNAs, which may be part of protein—RNA complexes.

Generally, investigators compare two or more sets of cells that are grown and treated in parallel. One might think that since these various incidental experimental parameters are held constant in the course of a single experiment, they will not alter the difference between the control and experimental groups. However, if the baselines change dramatically when the incidental parameters are altered (and they often do), this will also change the apparent magnitude of any group differences that are detected. For example, suppose that cells are induced to synthesize 10 units of new protein X when they are exposed to Interleukin-6. Then, the difference will be $10 + X/X$, where X is the amount of protein X present in the control group. **If the baseline content of X is 10 units, this will produce a doubling or twofold increase** $(= (10 + 10) /10 = 2)$, which is a relatively large proportional increase. In contrast, **if the baseline content of protein X is 100 units, the same response will only produce a 10% increase** $(= (10 + 100) / 100 = 1.10)$.

Cultured cells are popular because they are inexpensive, easy to experiment on, and easy to elicit responses that are dramatic and that differ across experimental groups. Nevertheless, most articles that describe cell culture experiments cannot be replicated by others, in part because the authors do not describe precisely all of the relevant experimental and culture parameters involved in the experiments, and in part because the same cells are not frozen and archived (so that they are not available for restudy). Nor are the findings demonstrated to be robust in most cases, because the findings are not often compared systematically across a series of different experimental conditions, as part of the same article.

Tip: At a minimum, a robust experiment is likely to compare several cell lines in parallel, carry out dose—response curves, and follow time courses of responses at multiple time points. As well, the experiments should be repeated several times independently before being reported.

A. DESIGNING YOUR EXPERIMENT

Animal Behavioral Experiments

Mice and rats are widely used in nearly all areas of biology and medicine. Here, I will consider how they are reported for behavioral studies in psychology, neuroscience, and psychiatry. These animals are genetically well characterized and are kept in elaborately well-controlled, quarantined animal facilities. Thousands of published articles have appeared over decades utilizing assays that, at first glance, seem to be relatively standardized. For example, "anxiety" in these animals is often measured using the so-called elevated-plus maze (Fig. 1.3), whereas "depression" is measured using the forced swim test, and "memory" is commonly measured using the Morris water maze.

Yet it has been well documented that the behavioral responses of mice and rats vary widely among strains (both inbred and outbred strains). Certain strains are preferentially chosen or avoided for different purposes depending on their differing baseline "personality" traits and response tendencies. In fact, by comparing the responses as they vary across a large number of strains, which differ systematically in their genetic makeup, one can identify specific genes that contribute to behavioral responses [6]. Even among animals of the same strain, their behavior in a given experiment may depend on the commercial supplier (and even from which warehouse the animals are obtained!) as well as an animal's age, gender,

FIGURE 1.3 A plus maze that is used to assess the willingness of a mouse or rat to walk out on a lighted or exposed platform. *By Ywlliane—Own work, CC BY-SA 3.0. https://commons. wikimedia.org/w/index.php? curid=49554753.*

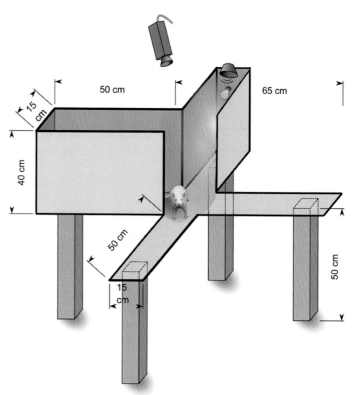

estrus stage, and birth order. Stressful handling is an important variable, which is encountered both at the time of the experiment and stresses encountered during early life. Moreover, responses may be different depending on which experimenter is carrying out the assays, the time of day, and even whether the experimenter handling the animals is female or male [7,8]. An important variable is whether the animal facility also houses animals that are predators (e.g., rats are predators for mice, and cats and dogs are predators for rats). In fact, Pavlov (famous for his dogs that salivated when they heard a bell) had his own building constructed to minimize such variables.

Crabbe et al. carried out a large-scale attempt to reproduce a battery of six behaviors in several inbred strains and one null mutant by simultaneous testing in three laboratories, keeping as many experimental parameters fixed as possible, and holding the behavioral assays as standardized as possible. Not only were there substantial differences between laboratories in baselines, both quantitatively and qualitatively, but also interstrain differences that varied markedly across laboratories [9].

Richter et al. tested whether *increasing* natural variation in each group might increase reliability of findings: Six laboratories independently ordered 64 female mice of two inbred strains (C57BL/6NCrl and DBA/2NCrl) and examined them for strain differences in five commonly used behavioral tests [10]. Instead of holding conditions as constant as possible, they systematically varied the animals' age and cage enrichment within each group. This modified design increased reproducibility somewhat but still did not substantially alter the fact that different results were reported across laboratories [10].

The problem is all the more serious in comparing experiments across laboratories that do not standardize their assays carefully. Despite the names, "standard" assays, such as the elevated-plus maze, actually have no standard way of scoring that applies uniformly in all cases (e.g., the assays need to be modified ad hoc for animals with different motor abilities and different body sizes).

In practice, investigators simply carry out a manipulation on two or more groups and report whether they observe a statistically significant difference in behavioral assay scores. But as I emphasized in the case of cell culture experiments, observing a statistically significant difference is not enough; it is not even a primary experimental outcome! The magnitude of differences observed between groups depends on the baseline assay values themselves, which vary exquisitely with minor variations in laboratory conditions and hence are generally not robust. The current state of affairs is that baseline scores in the control group are not necessarily comparable or similar across different published articles. A score on the elevated-plus maze test reported in one paper cannot be directly compared with a score in a different paper, even for the same strain of mice. This is like using a scale without zeroing it or calibrating it!

In summary, tens of thousands, perhaps hundreds of thousands, of peer-reviewed articles have been published, in leading journals and from leading research institutes, which have employed behavioral testing in mice and rats. Most, if not all, of these studies employed control experiments, and they observed differences across experimental groups that achieved statistical significance. These articles may indeed be valid statements of what was observed in a particular place, using a particular shipment of mice, encountering a particular dose of drug, and tested on a particular apparatus at a particular time. Yet, I hope I have convinced you that as a whole, this field fails basic standards of reproducibility and robustness.

A. DESIGNING YOUR EXPERIMENT

THE WAY FORWARD

If things are so bad (and they are), how then does science seem to move forward and apparently at such an accelerating rate? Articles, even those that are individually not reproducible or robust, do have a certain undeniable scientific value, for example, in pointing out promising new research directions. In the 1990s, a series of articles reported that the drug ketamine has antidepressant-like effects in mice and rats subjected to manipulations designed to elicit depression-like behaviors. In the 2000s this was followed up in humans in clinical studies [11] and randomized clinical trials [12]. Finally, in the current decade, both human and animal studies are continuing to investigate the underlying mechanisms and to establish whether this can be turned into a practical therapy for treating patients who have major depression. (If you are wondering why it has taken so long for ketamine to make it from the animal lab to the bedside, this is because ketamine has major neurological and psychiatric side effects of its own. Fortunately, recent research suggests that the antidepressant effects of ketamine can be separated from these side effects [13].)

Published articles can provide substantial value, particularly in the aggregate, when a number of independent experiments are collected, analyzed, and reviewed together (see Chapter 10). However, the current practice of publishing is very noisy and haphazard, and most importantly, trends can be discerned only AFTER scientists decide that a seminal finding is worth following up in the first place. **That decision can be made more rapidly and intelligently if the article that describes the initial finding has included cross-checks to demonstrate robustness.** In the following chapters, I will teach you how to design experiments that not only satisfy internal validity (control experiments and such) but also how to design them in a way that achieves reproducibility and robustness.

References

[1] Scoville WB, Milner B. Loss of recent memory after bilateral hippocampal lesions. J Neurol Neurosurg Psychiatry February 1957;20(1):11−21.

[2] Squire LR. The legacy of patient H.M. for neuroscience. Neuron Janurary 15, 2009;61(1):6−9. http://dx.doi.org/10.1016/j.neuron.2008.12.023.

[3] Gartler SM. Apparent Hela cell contamination of human heteroploid cell lines. Nature February 24, 1968;217(5130):750−1.

[4] Nelson-Rees WA, Daniels DW, Flandermeyer RR. Cross-contamination of cells in culture. Science April 24, 1981;212(4493):446−52.

[5] Frattini A, Fabbri M, Valli R, De Paoli E, Montalbano G, Gribaldo L, Pasquali F, Maserati E. High variability of genomic instability and gene expression profiling in different HeLa clones. Sci Rep October 20, 2015;5:15377. http://dx.doi.org/10.1038/srep15377.

[6] Philip VM, Duvvuru S, Gomero B, Ansah TA, Blaha CD, Cook MN, Hamre KM, Lariviere WR, Matthews DB, Mittleman G, Goldowitz D, Chesler EJ. High-throughput behavioral phenotyping in the expanded panel of BXD recombinant inbred strains. Genes Brain Behav March 1, 2010;9(2):129−59. http://dx.doi.org/10.1111/j.1601-183X.2009.00540.x.

[7] Chesler EJ, Wilson SG, Lariviere WR, Rodriguez-Zas SL, Mogil JS. Influences of laboratory environment on behavior. Nat Neurosci November 2002;5(11):1101−2.

[8] Sorge RE, Martin LJ, Isbester KA, Sotocinal SG, Rosen S, Tuttle AH, Wieskopf JS, Acland EL, Dokova A, Kadoura B, Leger P, Mapplebeck JC, McPhail M, Delaney A, Wigerblad G, Schumann AP, Quinn T, Frasnelli J, Svensson CI, Sternberg WF, Mogil JS. Olfactory exposure to males, including men, causes stress and related analgesia in rodents. Nat Methods June 2014;11(6):629−32. http://dx.doi.org/10.1038/nmeth.2935.

[9] Crabbe JC, Wahlsten D, Dudek BC. Genetics of mouse behavior: interactions with laboratory environment. Science June 4, 1999;284(5420):1670–2.

[10] Richter SH, Garner JP, Zipser B, Lewejohann L, Sachser N, Touma C, Schindler B, Chourbaji S, Brandwein C, Gass P, van Stipdonk N, van der Harst J, Spruijt B, Võikar V, Wolfer DP, Würbel H. Effect of population heterogenization on the reproducibility of mouse behavior: a multi-laboratory study. PLoS One January 31, 2011;6(1):e16461. http://dx.doi.org/10.1371/journal.pone.0016461.

[11] Berman RM, Cappiello A, Anand A, Oren DA, Heninger GR, Charney DS, Krystal JH. Antidepressant effects of ketamine in depressed patients. Biol Psychiatry February 15, 2000;47(4):351–4.

[12] Zarate Jr CA, Singh JB, Carlson PJ, Brutsche NE, Ameli R, Luckenbaugh DA, Charney DS, Manji HK. A randomized trial of an N-methyl-D-aspartate antagonist in treatment-resistant major depression. Arch Gen Psychiatry August 2006;63(8):856–64.

[13] Zanos P, Moaddel R, Morris PJ, Georgiou P, Fischell J, Elmer GI, Alkondon M, Yuan P, Pribut HJ, Singh NS, Dossou KS, Fang Y, Huang XP, Mayo CL, Wainer IW, Albuquerque EX, Thompson SM, Thomas CJ, Zarate Jr CA, Gould TD. NMDAR inhibition-independent antidepressant actions of ketamine metabolites. Nature May 4, 2016;533(7604):481–6. http://dx.doi.org/10.1038/nature17998.

Choosing a Research Problem

INTRODUCTION

I cannot tell you what kind of scientific studies to pursue, any more than I can tell you what kind of clothes to wear, who to date, or which candidate deserves your vote. These decisions involve weighing criteria that are subjective and differ for each person. Still, making a bad (or at least suboptimal) decision can have far-reaching consequences. Not everyone looks good in a purple jumpsuit! So, in this chapter, let us celebrate the diversity of scientific investigations and experimental designs, while at the same time thinking about how to decide which experiment(s) are best suited to us in a given time and place.

Peer-reviewed scientific articles follow diverse formats: Some papers simply report single, self-contained observations. Others report a controlled experiment or a successive series of experiments that are assembled to tell a coherent story, suggest a particular model, or test a preexisting hypothesis. Some papers are explicitly written as letters to the scientific community. Some raise methodological issues or concerns. Others are lengthy reviews that summarize current knowledge and synthesize new concepts or propose new research directions. Experimental research articles can be, for example, observational, epidemiological, retrospective, or prospective. Some experiments are designed to be exploratory or pilot studies, whereas others are intended to be formal, definitive, comprehensive studies (e.g., randomized, placebo-controlled, double-blind clinical trials). All of these types of articles, and all of these types of investigations, can make valuable contributions to science.

Philosophers of science like to contrast **hypothesis-driven** versus **question-driven** versus **data-driven** research [1–3], or **applied** versus **methodological** versus **mechanistic** studies. Similarly, people who study ways to measure the impact of science (the field of scientometrics) spend a lot of effort trying to recognize instances of **incremental** progress versus **breakthroughs** or **transformative** science. Everyone would like to choose a line of work that will result in a breakthrough, but (like predicting the stock market) this is not something that is entirely under the control of the scientist—even a smart one. I suspect that many, if not most breakthroughs are apparent only in retrospect. Often, a breakthrough depends on the publication of supporting or complementary findings that appear only *after* the original article is published. Kary Mullis' invention of the polymerase chain reaction (PCR) was beautiful and ingenious but only became a widespread technique after other

Data Literacy
http://dx.doi.org/10.1016/B978-0-12-811306-6.00002-6

17

scientists later (and independently) characterized thermophilic bacteria and purified heat-tolerant Taq polymerase, which allowed repeated cycling to take place without the need for manual handling at each step [4].

SCIENTIFIC STYLES AND MIND-SETS

Overall scientific styles and mind-sets vary considerably, often across disciplines: For example, what we might call an **engineering** approach is to simulate the overt behavior of a system, regardless of how it does so. **An airplane is meant to achieve flight, not to imitate birds**. This is in contrast to those who would seek to understand the underlying components, interactions, and mechanisms that underlie flight. It is not only engineers who have an engineering attitude; for example, B.F. Skinner and his followers in psychology advocated a **black box analysis** of behavior, in which the study of the brain's responses and rewards was used to derive rules of behavior, regardless of the neural circuitry that generates them.

A **medical** mind-set tends to employ **case-based reasoning**, employing real-world knowledge, and **conditional probabilities**. An old saying passed around in medical school is: "When you hear hoof beats, expect horses, not zebras" (although the version I learned went, "but always check for zebras"!) (Fig. 2.1).

This empirical attitude is in contrast to a formal or **physics** approach, which is concerned with abstracting observations into universal scientific **laws**. Often physicists study toy models that are grossly oversimplified but are intended to capture certain key aspects of a phenomenon.

Biology researchers differ yet again. They often think in terms of establishing **cause** and **effect**, identifying **regulators** that drive a system versus **modulators** that affect a system more weakly, and understanding how events are shaped by genes and environment. Biologists are prone to think in terms of **generalizations** and **exceptions**. For example, the Central

FIGURE 2.1   **Dr. House is different—he always expects zebras!**

Dogma of Molecular Biology states that "DNA makes RNA makes protein," which is taught to all biology students even though hefty asterisks are attached [some exceptions include (1) reverse transcription of RNA into DNA, (2) epigenetic modifications of DNA that alter gene expression without altering DNA sequence, (3) noncoding RNAs that are functional without encoding any protein products, and so on].

Data scientists (those in computer science, operations research, statistics, and informatics) carry out large-scale, comprehensive data—driven analysis of patterns in large data sets, an effort sometimes called **data mining**. They often construct and analyze **networks** of interacting entities or combine disparate types of evidence to create new **data objects** (e.g., a "**pathway**" is a data object inferred from a set of interacting proteins).

PROGRAMMATIC SCIENCE VERSUS LILY-PAD SCIENCE

At an even larger perspective, scientists differ in how they think about problems: Some think about the entities that they study in purely conceptual terms, whereas others visualize the entities as having size, shape, charge, grooves, and even tastes and odors. Ask a scientist to doodle their problem—do they write down words and sentences, or do they draw a picture or diagram?

Moreover, scientists differ in terms of how big they think! I do not necessarily mean THINK BIG like going to the moon or curing cancer. Rather, I mean that some scientists (particularly students) think only in terms of one experiment at a time, or one publication at a time. Others are guided by **research programs** that are often explicitly goal directed (e.g., identify the nucleotide sequences of the human genome, or identify better ways of identifying which patients are likely to respond to a given antidepressant drug). Since research grants are generally 3—10 years in duration, **programmatic science** is often envisioned and designed to be achievable within that time frame.

An alternative to this is **lily-pad science**, in which the scientist carries out one experiment and asks what new, interesting, or especially important problem is raised by it, then leaps to investigate that new problem even if it belongs to a different scientific domain or discipline altogether. Pasteur is perhaps the best known example of a scientist who thrived on this approach: He was hired to investigate why beer was becoming spoiled in a brewery, and when examining the dregs of the fermentation tanks he discovered that certain crystals have asymmetric shapes that twist only to the left or to the right. He hypothesized that the left-turning crystals were associated with living organisms, which led to his studies of fermentation, the germ theory, the debunking of spontaneous generation of life, and eventually studies of rabies vaccination—moving from industry to geology to biology to medicine along the way.

At the highest level of abstraction, scientific problems are often viewed in **paradigms or world views** that specify the types of problems that are considered amenable to scientific study in the first place, the way that these problems are tackled and the types of evidence that are considered to be permissible. Subjective feelings (phenomenology), beauty (aesthetics), consciousness, and the placebo effect are examples of topics that have gone in and out of fashion over the years as being appropriate (or not) for scientific study. Paradigms often are associated with events or movements that encapsulate an entire century. For example, if the 19th century was characterized by Darwinism and scientists worked as **individuals**, the 20th century was arguably the century of the bomb—and the focused **Big Science** approach that led to the development of the atomic bomb also led, postwar, to other focused Big Science **team** efforts such as high energy physics, the war on cancer, and the Human Genome Project. So far, it appears that the 21st century may become known as the century of Big Data, characterized by networks of computers, and interventional experiments may be at least partially replaced by data mining analyses.

CRITERIA (AND MYTHS) IN CHOOSING A RESEARCH PROBLEM

Is there any winning formula or recipe for doing successful science? One can find many autobiographical writings in which famous scientists give the secret to their success. The "secrets" are as varied as the scientists themselves [5—12]. Although these memoirs are fascinating to read, my own take-home message from this genre is that you should create your own life and not attempt to pattern yourself after any other person at all! As William Blake [13] said, "One Law for the Lion and Ox is Oppression." No single scientific path will be suited for math wizards, tinkerers, and daydreaming artists.

Learning how to choose scientific directions and write papers remains an art, learned by apprenticeship. One must consider practical issues—for example, to name only a few—is the topic an extension of your current work, or a totally new direction? Is it feasible to tackle with existing personnel, equipment, budget, and methods? Will the proposed experiment have the power to give a clear-cut, definitive yes/no answer? One must also consider the "interestingness" issues—again, to name only a few—is the hypothesis or proposed model beautiful? If true, will it solve a major problem or open up a new field of investigation? Will the findings generalize to other problems or domains?

Myths

Besides identifying positive attributes of good projects, I have noticed that many students avoid certain types of otherwise promising research problems because of naïve beliefs and attitudes that are worth discussing—and debunking—in some detail. For example, should a research question be logical? Should you avoid a research question that directly contradicts reliable experimental findings? Should your experiment be able to conclusively test or disprove your hypothesis? Should a good experiment produce findings that remain "true" over time? The naïve answer to each of these questions is yes. But the truth is more interesting and revealing.

Should a Research Question Be Logical?

Some type of logic is generally a good thing in planning experiments, but deductive logic can only take you so far. For example, one of the experiments that I was involved in was seemingly based on solid logical reasoning [14]: We read reports that the antidepressant fluoxetine stimulates a protein called BDNF in brain, and separately, we saw reports that BDNF stimulates neurogenesis (the production of new neurons) in the adult dentate gyrus.

"Therefore" one might logically hypothesize that fluoxetine stimulates neurogenesis, right? Yet this reasoning was not entirely compelling, since it is not clear whether fluoxetine stimulates BDNF (1) to the right level, (2) for the right amount of time, or (3) in precisely the right place in the brain. Furthermore, it was certainly possible that the effects of fluoxetine might be restricted to certain species or ages, or might act only briefly, or might be overshadowed by other effects that produce opposing actions to decrease neurogenesis. So deductive logic had enough strength to encourage us to carry out the experiments but not enough strength to make us sure of the outcome!

More often, scientific discovery involves intuitive leaps and follows leads according to rules that are, simply put, nonlogical. For example, the classic discovery of lithium as a treatment for bipolar disease started out logically enough but then took a sudden right turn (see Box 2.1).

Tip: Another powerful strategy is to work backward from the desired result. Ask not, "what is the likelihood that a hypothesis is true?" But rather, "if the hypothesis is true, what is the bang for the buck?"

For example, in 1996, I had joined the research group of Erminio Costa, an eminent molecular pharmacologist who in his 70s had embarked on a final chapter of his career and was

---

BOX 2.1

THE DISCOVERY OF LITHIUM AS A TREATMENT FOR BIPOLAR DISEASE BY JOHN CADE

John Cade hypothesized mania to be a "state of intoxication of a normal product of the body circulating in excess" [15]. To find a toxic agent in the urine of manic patients, he injected guinea pigs with the urine of people with mania versus normal controls. The urine of manic patients was particularly toxic, killing animals at much lower dosages than urine from controls. Cade then tested different constituents of urine to identify the lethal compound, finding that urea led to exactly the same mode of death as urine. However, manic patients did not have higher concentrations of urea. Thus, he began to search for substances that modify the toxic effect of urea. Uric acid enhanced the toxicity of urea, but because uric acid is insoluble in water, he injected urea into the guinea pigs together with the most soluble urate, lithium urate. The combined toxicity was much less than expected, suggesting that lithium may be protective.

But now Cade takes a leap: Injecting the animals with lithium carbonate only, he found them to become lethargic and unresponsive. Notice that he has now gotten rid of manic patients, urine, urea, and even uric acid and is simply injecting normal guinea pigs with a normal lithium salt. However, seeing how the guinea pigs have become calm, he wonders if lithium might make manic patients calm too! Despite the total lack of linear logic in this speculation, it turns out that lithium does make manic patients calm—and it does so without affecting healthy people! Indeed even today, lithium remains one of the true wonder drugs of psychiatry [15].

A. DESIGNING YOUR EXPERIMENT

looking for a new approach to understanding schizophrenia. There were certainly indications that schizophrenia reflects abnormal brain development and that the cerebral cortex (particularly prefrontal cortex) is particularly involved in the type of higher level functions that are altered in schizophrenia. Scientists would like to identify an animal model of schizophrenia that shares not only behavioral similarities to the human disease but also shares some of the critical developmental and molecular defects as well. The way forward was not clear. So instead, Dr. Costa took the way backward!

At this time, reelin, the gene responsible for the *reeler* mutation in mouse, had just been identified as encoding an extracellular matrix protein produced only by specialized cells at the roof of the developing cortex [16]. *Reeler* mice show abnormal layering of neurons in the cortex, and though they are surprisingly intact in their behavioral repertoire, Dr. Costa had observed them and thought that they "act weird." There was no evidence, at the time, that reelin should be involved in schizophrenia. Yet, reasoning backward, *if* reelin should prove to be involved in schizophrenia, a gene would now be identified that regulates cortical development in a well-defined manner, and which already has an animal model that is amenable to mechanistic study. One could immediately ask specific questions and get clean yes/no answers: Is reelin expressed in human brain? Is it only found during development, as expected, or can it still be detected in the mature brain? Are there any reelin deficits observable in postmortem brain samples of humans dying with a diagnosis of schizophrenia? Costa's group found that the reelin gene is indeed expressed in mature human cortex, and the reelin levels measured in schizophrenics were only half of matched nonpsychiatric control subjects [17]. The observed reelin deficit remains, to date, as one of the most reliably replicated postmortem tissue findings in the schizophrenia field.

Should You Avoid a Research Question That Directly Contradicts Reliable Experimental Findings?

Contradictions are, in fact, routine. In the case of reelin, Costa had to ignore the early expectations that reelin would only be measurable in the early developing brain. Furthermore, as part of my contribution to the reelin project, I sought a tissue source that could be readily sampled to measure reelin in living humans. Early reports implied that reelin is a brain-specific protein, but I reasoned that (1) reelin has the structural features of an extracellular matrix protein and (2) many, if not all, extracellular matrix proteins are expressed in the blood. So it made sense to look for reelin in blood [18].

Another classic example is the discovery of nerve growth factor (NGF) by Rita Levi-Montalcini, for which she eventually won the Nobel Prize. At the time, she hypothesized that NGF would be specific to tumors and not secreted by normal tissues. However, the normal tissue that she chose as a negative control had similar growth-promoting activity as the tumor. Instead of abandoning her project because of this apparent contradiction, she pressed forward anyway and isolated NGF as a specific protein [8].

The take-home lesson: Science does not simply advance on the shoulders of others, because at the outset much of what we believe to be true is not so, and prior-published experiments that seem to be quite definitive may not be (or, more often, they are only true in certain contexts). Scientists should expect uncertainty and should be comfortable operating in an uncertain environment.

STRONG INFERENCE

Up to this point, we have used the terms "**research question**" and "**hypothesis**" pretty interchangeably. However, they are not quite the same thing. A research question is a much more general, open-ended way of creating a scientific study, which does not presuppose the answer—for example, "what percent of US children are living below the poverty line?" In contrast, a hypothesis represents a guess or prediction that the experiment ought to test conclusively with a yes/no answer—for example, "Do 20% or more US children live below the poverty line?"

Having a hypothesis is considered desirable, because it encourages you to design experiments that satisfy **Strong Inference** [19]. Strong Inference designs an experiment so as to test an outcome that would ONLY be predicted by one hypothesis and would not plausibly be expected to occur according to any other hypothesis that can be formulated or envisioned by the investigators. This is a great mental exercise for scientists who are planning their experiments. It helps sharpen the question in two ways: First, it reminds the scientist that, if possible, they should design an experiment that produces a clean yes/no answer. Second, it forces one to ask, "what ARE all the other alternative hypotheses that I can think of, and what outcomes would I expect them to produce?"

Unfortunately, Strong Inference can only be achieved in arenas that are sufficiently advanced, both conceptually and in terms of experimental reagents and models, to test a specific outcome. For example, if one is interested in the role of a particular sodium channel in a particular neuron, one might be able to predict critically what happens if a serine residue at position 112 is mutated to an alanine. But this depends on having a lot of prior knowledge and expertise: the sodium channel's protein sequence; knowledge that the serine is phosphorylated in vivo; prior investigations of other types of channels; availability of methods to mutate specific residues; and ability to measure the permeability of that channel within the context of that neuron.

DESIGNING STUDIES AS COURT TRIALS

A broader, much more widely applicable version of Strong Inference is to regard **a scientific study as a court trial**. That is, **the job of the scientist is to persuade his or her peers in the scientific community beyond a reasonable doubt**. Some evidence is circumstantial, some is hearsay, some is correlational; certainly, finding a smoking gun is the most persuasive form of evidence but is not always present. **Convincing a jury of your peers beyond a reasonable doubt means that all of the alternative explanations that they can think of seem far less likely than the one you are proposing.** Conducting science in this manner keeps scientists on their toes in the planning, execution, and reporting of experiments and can be very broadly applied indeed. But remember that convincing a jury of your **findings** (i.e., the validity of the data and effects that you are reporting) and convincing them of your preferred **interpretation** of these findings are two entirely separate things!

What *the Sixth Sense* Teaches Us About Scientific Evidence and Persuasion [20]

The science fiction movie *The Sixth Sense* serves as a platform for discussing attitudes that are helpful for scientific investigation, such as "keep an open mind," "reality is much stranger than you can imagine," and "our conclusions are always provisional at best." It is also a great metaphor to illustrate how scientists use evidence to persuade the scientific community.

When Haley Joel Osment (Fig. 2.2) says, "I see dead people," does he actually see ghosts? Or is he hallucinating? It is important to emphasize that these are not merely different viewpoints, or different ways of defining terms. If we argued about which mountain is higher, Everest or K2, we might disagree about which kind of evidence is more reliable, but we would fundamentally agree on the notion of measurement. By contrast, in *The Sixth Sense*, the same evidence used by one **paradigm** to support its assertion is used with equal strength by the other paradigm as evidence in *its* favor. In the movie, Bruce Willis plays a psychologist who assumes that Osment must be a troubled youth. However, the fact that he says he sees ghosts is also evidence in favor of the existence of ghosts (if you do not reject out of hand the possibility of their existence).

These two explanations are **incommensurate**. One cannot simply weigh all of the evidence because each side rejects the type of evidence that the other side accepts, and regards the alternative explanation not merely as wrong but as ridiculous or nonsensical. It is in this sense that a paradigm represents a failure of imagination (Fig. 2.3)—each side cannot imagine that the other explanation could possibly be true, or at least, plausible enough to warrant serious consideration. The failure of imagination means that each side fails to notice or to seek objective evidence that would favor one explanation over the other. For example, during the episodes when Osment saw ghosts, the thermostat in the room fell precipitously and he could

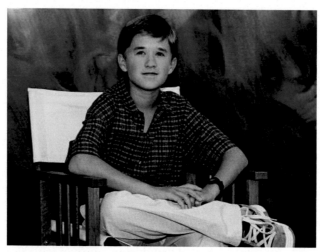

FIGURE 2.2 **Haley Joel Osment (born 1988) is an Academy Award–nominated American actor who came to fame with his starring role in "The Sixth Sense" (1999).** *By Thomas from Vienna, Austria—Haley Joel OSMENT, CC BY-SA 2.0, https://commons.wikimedia.org/w/index.php?curid=4905564.*

FIGURE 2.3 **An idea reimagining itself.** *Reprinted from Smalheiser NR. How many scientists does it take to change a paradigm? New ideas to explain scientific observations are everywhere—we just need to learn how to see them. EMBO Rep October 2013;14(10):861—865 with permission.*

see his own breath. This certainly would seem to constitute objective evidence to favor the ghost explanation, and the fact that his mother had noticed that the heating in her apartment was erratic suggests that the temperature change was actually happening. But the mother assumed that the problem was in the heating system—so the "objective" evidence certainly was not compelling or even suggestive on its own.

Osment did eventually convince his mother that he saw ghosts, and he did it in the same way that any scientist would convince his colleagues: Namely, **he produced evidence that made perfect sense in the context of one, and only one, explanation.** First, he told his mother a secret that he said her dead mother had told him. This secret was about an incident that had occurred before he was born, and presumably she had never spoken of it, so there was no obvious way that he could have learned about it. Next, he told her that the grandmother had heard her say "every day" when standing near her grave. Again, the mother had presumably visited the grave alone and had not told anyone about the visit or about what was said. So, the mother was eventually convinced that Osment must have spoken with the dead grandmother after all. No other explanation seemed to fit all the facts.

Is this the end of the story? We, the audience, realize that it is possible that Osment had merely guessed about the incidents, heard them secondhand from another relative or (as with professional psychics) might have retold his anecdotes while looking for validation from his mother. The evidence seems compelling only because these alternatives seem even less likely. It is in this same sense that when scientists reach a conclusion, it is merely a place to pause and rest for a moment, not a final destination. (Modified from Ref. [20] with permission.)

A. DESIGNING YOUR EXPERIMENT

Remember that even Strong Inference and court trials are not infallible guides to truth. One can rule out every alternative hypothesis that you can conceive of, and still not get the right answer—because often, the correct hypothesis is something not yet conceived of at the time [21]! More importantly, many fields of research are relatively primitive, to the degree that one cannot formulate a question precisely enough to pose a falsifiable hypothesis. Many of the scientific studies that lack Strong Inference, which are looked down on as "incremental" or "descriptive," are actually valiant efforts to chip away at very tough or vague problems. On the other hand, if possible, when designing exploratory experiments, it is good advice try to arrange things so that one will gain useful knowledge from the experiment regardless of outcome.

Tip: If an experiment is interpretable only when the outcome is positive and will be a complete waste of time if the outcome is negative, that is the opposite of Strong Inference. It is a red flag. Proceed at your own risk!

"Degrees of Freedom"

A major cause of limited robustness and reproducibility occurs when there are too many **"degrees of freedom"** in the experiment. [Note: This term, though commonly used in this context, should not be confused with the statistical term "degrees of freedom," which has a completely different meaning (see Chapter 3).] For example, a typical neuroscience experiment might ask: "is the amygdala required for recognition of siblings in postpubertal rats?" Although that sounds reasonable and meaningful, there is actually a huge amount of ambiguity in how the question is being posed—two people asking ostensibly the same question might pursue it in different ways and get totally different answers such as in the following cases:

1. What exactly, for example, is the amygdala? The borders of the amygdala are not entirely agreed on. Since the amygdala comprises a complex of 13 distinct nuclei (or more), its functional role in sibling recognition might depend on the action of all nuclei working as a whole, might involve some nuclei and not others, or it might depend on discrete interactions among nuclei. The anatomy and pattern of interconnections may vary across different strains of rats and in different species.
2. One way to test the question is to remove or silence the amygdala. How should it be done? Different methods of lesion (electrolytic, chemical, optogenetic silencing, etc.) will affect the amygdala and its neuronal and glial cells very differently. Stimulating the amygdala, or expressing some exogenous gene selectively in amygdaloid neurons, will involve the amygdala in a quite different fashion (and moreover, will possibly only involve particular subtypes of neurons). Each of these methods may affect the amygdala for different durations as well.
3. How does one define and measure "recognition of siblings"? By measuring changes in heart rate? Changes in gaze? Do the assays require the rat to have intact motor ability, motivation, short-term memory, or sense of smell, which might be affected by lesions or other manipulations to the amygdala? If so, changes in the assay might cause the investigator to falsely conclude that recognition is affected, when it is the underlying motor or sensory acuities instead.

4. Even apart from these potential **confounds**, the amygdala might affect recognition of siblings by one mechanism under one set of conditions and by a different mechanism under other conditions. Male versus female siblings might be recognized by somewhat different neural mechanisms. Conversely, male rats and female rats might employ different neural mechanisms for recognizing their siblings.

In summary, the apparently simple and scientific question: "Is the amygdala required for recognition of siblings in postpubertal rats?" might be true in several different ways and simultaneously false in several different ways! What we need is not any single yes/no answer but rather an ensemble of answers that, taken together, provides insight into the anatomy and physiology of the amygdala. Approaching the question via multiple approaches, and as measured by multiple assays, is the key to achieving this insight (and, as will be argued in Chapter 4, is one key to robustness of experimental design in general).

What to Do When No Strong Hypothesis Is Apparent?

Most compelling scientific problems are vague and mysterious and have no clear point of attack. Is immortality achievable? Or can we at least delay or reverse aging? Why do we spend one-third of our lives sleeping? Can robots be programmed to have emotions? Can we predict crimes before they happen? Such topics are valid scientific questions, yet they are addressed far more often in science fiction than in the published scientific literature. Several strategies may be helpful for taking a vague problem and constructing experiments that have clear yes/no answers and can help point the field toward more precise hypotheses.

For example, let us consider the problem of schizophrenia, which affects roughly 1% of the population worldwide, and which remains rather mysterious (certainly, as compared to diseases such as sickle cell anemia that are caused by single-gene point mutations).

1. One technique, as mentioned, is to **think backward from the desired outcome**. For example, "if schizophrenia were an infectious disease, then what predictions would follow, that could be tested, that would NOT be expected to occur otherwise?"
2. Another strategy is to **nibble around the edges of the problem**. For example, one might study the time course of emergence of symptoms in adolescents or examine natural variation in the expression of disease across different populations (urban vs. rural).
3. A popular approach is to **examine a simple model system** that exhibits some (usually not all) aspects of schizophrenia, for example, various mouse models of "schizophrenia" based on genetic manipulations or environmental insults, or humans who have a particular chromosomal abnormality (22q11.2 microdeletions [22]) who have greatly elevated risk of developing schizophrenia.
4. A particularly pragmatic endeavor is to **develop new methods** for studying the nervous system, in the hope that they may detect alterations missed by currently available methods. Often these methods open up new lines of investigation in other problem areas, which may eventually shed light on the original problem of interest. For example, who knows why the microelectrode, the patch clamp, or the MRI were first invented? Regardless of the original reasons, these methods have moved our knowledge of nervous function forward on a broad front.

A. DESIGNING YOUR EXPERIMENT

5. Another possible strategy is to carry out **data-driven studies**, for example, collecting a large number of features of human subjects with schizophrenia, other conditions and healthy controls, perhaps including genetic markers, protein biomarkers, prenatal history, diet, etc., and looking for statistically significant patterns that correlate with schizophrenia or with particular subtypes or outcomes.
6. One might also look for clues from **clinical case reports**, for example, cases where schizophrenic patients were treated for some other cooccurring disease such as asthma, where the treatment unexpectedly improved their psychotic symptoms.

Should a Good Experiment Produce Findings That Remain "True" Over Time?

There are those who believe that science marches steadily toward truth. In this view, the body of experimentation performed by the scientific community increases the total amount of knowledge over time, and if it does not always establish which facts are entirely correct, at least it reliably winnows out certain ideas and hypotheses as wrong or in error. These errors can and should be discarded on the trash pile of history.

Me? I see this trash pile as a valuable source for recycling!

Consider Einstein's theory of general relativity, which has certainly survived experimental testing and gained general acceptance. Should it not supercede and replace Newton's laws of motion? But Newton's laws are still regarded as corresponding to common sense, are universally taught in schools, and are generally used for computations. Einstein's theory is only taught in advanced specialized classes and often regarded as a "correction" to Newton's laws, which are needed only when velocities approach the speed of light.

Consider Lamarck's theory of inheritance of acquired characteristics, which proposed that environmental events could cause a directed response in organisms that is inherited and affects their subsequent generations. This proposal had been not only thoroughly discredited but also for decades was held up as a textbook example of a dead, refuted hypothesis [23]. Even though a variety of authors had provided arguments and experimental evidence in favor of Lamarckian phenomena over many years [23], mainstream scientists refused to take it seriously because Lamarck's proposal appeared to lack any known or even any conceivable mechanisms that could mediate it. The situation changed when it was found that double-stranded small RNAs can cause specific gene silencing that spreads within the body of *Caenorhabditis elegans* worms, including into the germ line, where the silencing affects their progeny for several generations [24,25]. Once the case occurring in *C. elegans* became well understood mechanistically, the transgenerational inheritance of environmental influences has rapidly become investigated and widely accepted in other organisms including bacteria and mammals.

Consider traditional Chinese medicine, which if not superceded by Western medicine, is at least ignored by Western medical schools and practitioners. Yet the 2015 Nobel Prize for Medicine was shared by Youyou Tu (Fig. 2.4). She studied ancient procedures that are used to prepare folk remedies for malaria from sweet wormwood in traditional Chinese herbal medicine, and followed their empirical guidelines to isolate a specific antimalarial compound, artemisinin.

Consider the experiment mentioned earlier reporting that fluoxetine, an antidepressant, stimulates neurogenesis in the adult dentate gyrus. The original experiments have held up

FIGURE 2.4   **Tu Youyou, Nobel Laureate in medicine, Stockholm, December 2015.** *By Bengt Nyman—Own work, CC BY-SA 4.0, https://commons.wikimedia.org/w/index.php?curid=45490782.*

extremely well over time and have been replicated many times [26]. The finding almost immediately stimulated several bolder hypotheses: (1) antidepressant drugs generally exert their therapeutic effects by stimulating neurogenesis; (2) stimulation of neurogenesis is a prerequisite for the beneficial effects of any antidepressant agent; and (3) the disease of depression itself may be caused by a lack of ongoing neurogenesis. Yet after almost 15 years of sustained research activity, and hundreds of published papers, it is still not clear exactly how neurogenesis and depression are related, and none of these hypotheses may turn out to be true. Yet most investigators in the field would still say that this has been a highly fertile area of investigation [26], because it has led to a much deeper understanding of the role of neurogenesis in learning and other normal brain functions and to a much better understanding of how antidepressant drugs affect the brain.

   Tip: An experiment that opens new doors of investigation is a good experiment—even if it is not immediately clear where it will lead.

References

[1] Kell DB, Oliver SG. Here is the evidence, now what is the hypothesis? The complementary roles of inductive and hypothesis-driven science in the post-genomic era. Bioessays January 1, 2004;26(1):99—105.

[2] Nabel GJ. Philosophy of science. The coordinates of truth. Science October 2, 2009;326(5949):53—4.

[3] Glass DJ, Hall N. A brief history of the hypothesis. Cell August 8, 2008;134(3):378—81.

[4] Fore Jr J, Wiechers IR, Cook-Deegan R. The effects of business practices, licensing, and intellectual property on development and dissemination of the polymerase chain reaction: case study. J Biomed Discov Collab July 3, 2006;1:7.

[5] Silva AJ. The science of research: the principles underlying the discovery of cognitive and other biological mechanisms. J Physiol Paris July—November 2007;101(4—6):203—13.

[6] Snyder SH. The audacity principle in science. Proc Am Philos Soc June 1, 2005;149(2):141–58.

[7] Ramon y Cajal SR. Advice for a young investigator. MIT Press; 2004.

[8] Levi-Montalcini R. In praise of imperfection: my life and work. In: Alfred P. Sloan foundation series. New York: Basic Books; 1988c. 1988.

[9] Medawar PB. Advice to a young scientist. Basic Books; 2008.

[10] Watson J. The double helix. 2012. Hachette, UK.

[11] Alon U. How to choose a good scientific problem. Mol Cell September 24, 2009;35(6):726–8.

[12] Kaiser JF. Richard hamming-you and your research. Berlin, Heidelberg. In: Simula Research Laboratory, Springer; 2010. p. 37–60.

[13] Blake W. The marriage of heaven and hell. 1790–1793. Poetic Works.

[14] Manev H, Uz T, Smalheiser NR, Manev R. Antidepressants alter cell proliferation in the adult brain in vivo and in neural cultures in vitro. Eur J Pharmacol January 5, 2001;411(1–2):67–70.

[15] Mitchell PB. On the 50th anniversary of John Cade's discovery of the anti-manic effect of lithium. Aust N Z J Psychiatry October 1999;33(5):623–8.

[16] D'Arcangelo G, Miao GG, Chen SC, Soares HD, Morgan JI, Curran T. A protein related to extracellular matrix proteins deleted in the mouse mutant reeler. Nature April 20, 1995;374(6524):719–23.

[17] Impagnatiello F, Guidotti AR, Pesold C, Dwivedi Y, Caruncho H, Pisu MG, Uzunov DP, Smalheiser NR, Davis JM, Pandey GN, Pappas GD, Tueting P, Sharma RP, Costa E. A decrease of reelin expression as a putative vulnerability factor in schizophrenia. Proc Natl Acad Sci USA December 22, 1998;95(26):15718–23.

[18] Smalheiser NR, Costa E, Guidotti A, Impagnatiello F, Auta J, Lacor P, Kriho V, Pappas GD. Expression of reelin in adult mammalian blood, liver, pituitary pars intermedia, and adrenal chromaffin cells. Proc Natl Acad Sci USA February 1, 2000;97(3):1281–6.

[19] Platt JR. Strong inference. Science 1964;146(3642):348–53.

[20] Smalheiser NR. How many scientists does it take to change a paradigm? New ideas to explain scientific observations are everywhere—we just need to learn how to see them. EMBO Rep October 2013;14(10):861–5.

[21] http://edge.org/panel/sendhil-mullainathan-what-big-data-means-for-social-science-headcon-13-part-i.

[22] Bassett AS, Chow EWC, Weksberg R. Chromosomal abnormalities and schizophrenia. Am J Med Genet 2000;97(1):45–51.

[23] Jablonka E, Lamb MJ, Avital E. 'Lamarckian' mechanisms in darwinian evolution. Trends Ecol Evol May 1998;13(5):206–10.

[24] Fire A, Xu S, Montgomery MK, Kostas SA, Driver SE, Mello CC. Potent and specific genetic interference by double-stranded RNA in *Caenorhabditis elegans*. Nature February 19, 1998;391(6669):806–11.

[25] Rechavi O, Houri-Ze'evi L, Anava S, Goh WS, Kerk SY, Hannon GJ, Hobert O. Starvation-induced transgenerational inheritance of small RNAs in *C. elegans*. Cell July 17, 2014;158(2):277–87.

[26] Petrik D, Lagace DC, Eisch AJ. The neurogenesis hypothesis of affective and anxiety disorders: are we mistaking the scaffolding for the building? Neuropharmacology January 2012;62(1):21–34.

A. DESIGNING YOUR EXPERIMENT

Many scientists, many ways of investigating.

Basics of Data and Data Distributions

INTRODUCTION

To discuss how to design and analyze experiments, we need to introduce a certain number of concepts and vocabulary terms. Let us choose a simple population that an investigator might be interested in, say, children (aged 0–18 years) living in the United States, and for simplicity, let us measure their heights in 2016. The heights are called **outcomes**, **endpoints**, or the **dependent variable**. What will the resulting distribution of heights look like? Certainly, all values are greater than zero, and there will be a **minimum** value (children are not likely to be 2 inches tall) and a **maximum** (children are not likely to be 7.5 ft tall), with most people falling between 3 and 6 ft. The span from the minimum to maximum value is called the **range**. Thus, the curve will have some sort of hump, although there is no reason to expect that it will necessarily resemble the **bell-shaped curve** known as the **normal distribution** (Fig. 3.1).

The overall population of children is very heterogeneous, and different subgroups of children may have very different height distributions. The third graders will be, on average, taller than the second graders (although there will be very substantial overlap between these two groups). A sociologist might seek to compare children from rich versus poor families, or Northerners versus Southerners, or urban versus rural families. Each such comparison results in two curves, or distributions, and statistical testing is used to determine whether the two distributions do or do not have significantly different height distributions.

AVERAGES

Perhaps the most common way to compare two distributions is to ask whether the "average" child from one group is taller than the "average" child from the other group. Several types of average value are commonly considered, each being preferred in different situations:

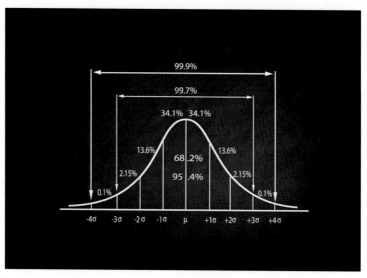

FIGURE 3.1   The normal distribution. This shows a continuous curve, centered at the mean value μ of a dataset, and falling away symmetrically to infinity in both directions. Real datapoints are unlikely to fall exactly on the curve; rather, this represents a best-fit idealization. The distribution is marked off in SD units.

The **arithmetic mean**, by far the most common type of mean used in statistics, is computed by adding up all of the values x_i and dividing by n, the number of datapoints in the population:

$$\text{Arithmetic mean} = (x_1 + x_2 + x_3 + \ldots x_n)/n$$

Unless stated otherwise, the **mean** usually refers to the arithmetic mean, and the vast majority of data studies employ it as standing for the average value. The arithmetic mean is a very simple, intuitive measure that is widely used for statistical inference, but can be influenced strongly by **outliers** (rare points with highly unusual values). Sometimes **truncated** or **trimmed** means are computed, which improve robustness by removing the extreme values at the low and high end of the distribution from consideration. Truncated means are not generally used as primary parameters, but as a cross-check on conclusions made using regular means.

The **geometric mean** or **geomean** is computed by multiplying all of the values together and taking the nth root:

$$\text{Geomean} = (x_1 \times x_2 \times x_3 \times \ldots x_n)^{-n}$$

The geomean is more appropriate than the arithmetic mean for handling distributions of datapoints that are measured on a **log scale** or that include points that are different from each other by **orders of magnitude**. Arithmetic means and geomeans can differ greatly from each other when distributions cover a wide range of values; note that the arithmetic mean of 10 and $1000 = 505$, whereas the geomean $= 100$. (Other types of means exist, notably the **harmonic mean**, but will not be covered here.)

Another very common type of average is the **median**, which defines the midpoint of the distribution—that is, half of the points are below the median, and half lie above. Both medians and means are widely used in statistics. The median is more robust than the mean value, in the sense that it is hardly affected by outliers. Computing the median involves assigning ranks or quantiles to the datapoints ordered from the lowest (rank 1) to the highest (rank n), and marking the median at the halfway point. Assigning ranks is straightforward except that often, a population has multiple points with the same value, or **ties**, which different schemes handle differently.

Ranking is also used to divide the population into **quartiles** (0%—25%, 25%—50%, 50%—75%, and 75%—100%) or **percentiles** (0%—1%, 1%—2%, and so on). Quartiles are useful for describing a distribution (Q1 = 25%, Q2 = 50%, and Q3 = 75%) and the 50% of points lying in the middle, between Q1 and Q3, are said to comprise the **Interquartile Range (IQR)**. The box plot (Fig. 3.2) is a way of plotting datapoints belonging to a group that indicates the position of the quartiles Q1, median (Q2), and Q3.

VARIABILITY

Another basic aspect of a data set is the degree of dispersion of values across the population. Are most of the points piled up near the mean value, or do they range over many orders of magnitude? First one calculates the **Sum of Squares (SS)**: you consider each datapoint and compute the difference between that value and the mean value μ. This difference is squared, so that the difference is always a positive number, and then the differences are added for all datapoints. In equation form:

$$SS = (x_1 - \mu)^2 + (x_2 - \mu)^2 + \ldots (x_n - \mu)^2$$

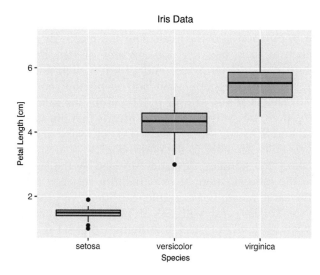

FIGURE 3.2 **Example of a graph employing box plots to summarize data distributions for each group.** Each box plot is framed by Q1 and Q3, with a thick line through it indicating the position of the median (Q2). Vertical lines, called "whiskers," are usually added to indicate values that lie above Q3 and below Q1. Note, however, that there is no standard way of displaying whiskers! Perhaps, most commonly the whiskers extend to values (1.5 × Q3) and (Q1/1.5) but sometimes whiskers show the entire range of the distribution (from maximum to minimum points), or 98% CI (from values achieved by only 2% of points, to values achieved by 98% of points). Points that fall outside the whiskers are usually shown as single dots, to display the full range of points including any outliers. *Courtesy of Araki Satoru, taken from https://commons. wikimedia.org/wiki/File:Simple-boxplot.svg with permission.*

A. DESIGNING YOUR EXPERIMENT

In short, **the parameter SS indicates the** *total* **amount of dispersion across the entire data set**, but to learn the average SS *per datapoint*, that is, the **variance** (Var), one divides SS by the number of datapoints in the population that are free to vary, i.e., the degrees of freedom (df). In formula form:

$$Var = SS/df$$

For a normal bell-shaped curve, df is simply N, the total number of points in the population, so that Var = SS/N. Other data distributions may involve datapoints that are subject to constraints, and not fully independent of each other, so that df may be less than N (see below).

The more familiar term **standard deviation (SD)** is the square root of the variance and simply states **the average amount by which each datapoint deviates from the mean value**. (Note that the variance states the *average amount squared* per datapoint, so taking the square root of the variance reverses the effect of squaring the differences in the first place.)

There are many cases where two distributions differ not in their means, but in their variances. In fact, a difference in variances may be the main outcome of an experiment, so it is important for investigators to consider and assess this possibility. For example, if two groups of children had the same mean heights, but differed greatly in variances, this would indicate that one group has greater heterogeneity. In turn, such a finding may stimulate further exploration to learn whether the greater dispersion of heights in one group reflects greater ethnic diversity, unequal access to food or health care, etc.

degrees of freedom (df): Consider a distribution that consists of only three points:

$$Mean = \mu = (x_1 + x_2 + x_3)/3$$

But the differences of each point from the mean must add up to zero!

$$(x_1 - \mu) + (x_2 - \mu) + (x_3 - \mu) = 0.$$

If you sample the three points from a population and then calculate the mean, df = 3. However, if the value for the mean is taken as fixed (i.e., it is the outcome of sampling), then given any two of the points, the third one is determined, and df = n−1 = 2, not 3. This situation occurs when taking a sample from the population and using the sample mean as an estimate of the population mean. That is, **if you sample n datapoints, the sample distribution does not have n degrees of freedom, but only n − 1. That means that the sample distribution has variance = SS/(n − 1)** and **SD** = $\sqrt{(SS/(n-1))}$. (Do not worry, this is probably the most difficult concept that you will encounter in the entire book.)

The **coefficient of variation (CoV)** is simply the standard deviation divided by the mean (CoV = SD/μ), and it is used to compare the intrinsic amount of variability between two different distributions that do not share the same mean. For example, consider two groups of people: one has a mean annual salary of $20,000/year with SD of 10,000, and another has a mean annual salary of $200,000/year, with SD of 10,000. Although the SD is the same, in the first distribution, 68.2% of the individuals will have incomes between $10,000 and $30,000, covering a threefold range. In contrast, in the second case, 68.2% of the individuals will have incomes between $190,000 and $210,000, covering a much smaller proportional difference (only 10.5%). The coefficient of variation in the first case is $10,000/20,000 = 0.5$, whereas in the second case it is $10,000/200,000 = 0.05$, a much smaller value that indicates its much smaller intrinsic variability.

THE BELL-SHAPED ("NORMAL") CURVE

The bell curve has many useful and amusing properties:

1. A true normal curve goes to infinity in both directions, although many curves that only take positive values may closely approximate a normal curve otherwise.
2. The point at 1 SD is situated at the inflection point of the curve, that is, where the convex part of the curve starts to become concave.
3. The mean ± 1 SD covers 68.2% (about two-thirds) of the area under the curve, or in other words, 68.2% of the datapoints fit within 1 SD from the mean. The IQR is a bit smaller than this, since it covers the middle 50% of points. (Actually, for a normal curve, one can use the IQR to estimate the SD, as SD $= {}^3/_4$ IQR, approximately.)
4. The point at 2 SD is also an important marker, since the interval encompassing the mean ± 2 SD covers 95% of the datapoints. Thus, the 95% CI runs from -2 SD to $+2$ SD. About 99.7% of the datapoints will fit within the mean ± 3 SD.
5. The rules just stated apply only to normal curves, not any arbitrary curve. However, regardless of the type of curve, Chebychev's Inequality states that at least 75% of the datapoints will fit within the interval from -2 SD to $+2$ SD.
6. As we said above, the degrees of freedom (df) of a bell curve are simply the number of datapoints N that make up the curve.
7. Why do so many real-life data sets follow the normal distribution, at least roughly? Those that do follow a linear model, in which each outcome x_i is partly determined by some fixed influence driving it to μ and partly determined by random fluctuations ε (or many small effects that act in both directions). In equation form,

$$x_i = \mu + \varepsilon$$

So, it is as if each datapoint is determined by starting with the value μ and then jiggling it up or down by some randomly chosen number ε. In some degree this can produce the distribution of heights in a classroom, the number of expected heads after flipping coins a very large number of times, and so on. Of course, most real-life data sets do not really follow a normal distribution closely, unless you squint at them. Those which have a single "hump" and are roughly symmetrical are called **quasinormal** and can often be treated statistically as if they were normal, as long as the number of datapoints per group is large enough (20–30 or more).

A. DESIGNING YOUR EXPERIMENT

NORMALIZATION

A nice use of the SD is to **normalize** a data distribution, so that different curves can be compared on the same footing, regardless of how many points they contain, or what actual values they express. For any datapoint, its distance from the mean μ can be stated in units of SD and expressed as its z-score. In formula terms,

$$z\text{-score} = (x_i - \mu)/SD$$

The 95% CI contains 95% of all points (for a bell curve, this will approximately lie within a z-score of -2 to $+2$). The 95% **CI** is an important parameter for analyzing and reporting experimental results.

The **standard form** for a data distribution is one that normalizes all values using z-scores, so that the mean $= 0$ and the SD $= 1$.

DISTRIBUTION SHAPE

Any data distribution is determined by three parameters: its mean, its variance, and its shape. **Skew**, the most important shape parameter for our purposes, is a term that refers to asymmetric curves. A normal curve is symmetric and has no skew. However, many phenomena are described by highly skewed distributions. For example, the **power law** distribution (Fig. 3.3) is given by $f(x) = x^{-k}$, where k is some positive constant. The power law arises in diverse situations: If you plot the size of cities against their number, you will see that a few cities have millions of residents, whereas most cities have only a few thousand or even fewer. Among published articles, a few are cited tens of thousands of times, whereas most are cited once or not at all. A few families are worth billions of dollars, whereas most have little net worth. A few words are extremely common, whereas most are rarely used. And so on. Often

FIGURE 3.3 **Example of a power law distribution.** This idealized plot makes the point that most people write few scientific articles, whereas a few write more than a thousand articles during their careers. *Courtesy of Phil Inje Chang, https://commons. wikimedia.org/wiki/File:PowerLawDistribution. png with permission.*

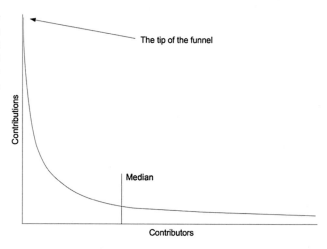

data follow the power law closely in a certain range of values but deviate at very high and low values due to practical constraints. For example, city sizes are limited by the fact that it is logistically difficult to engineer a city to accommodate 100 million people and their needs for food, water, and transportation. An interesting feature of a power law curve is that it follows a straight line when plotted on a log–log scale.

A PEEK AHEAD AT SAMPLING, EFFECT SIZES, AND STATISTICAL SIGNIFICANCE

In modern science, it is sometimes possible to observe and study an entire population of interest. We might have the entire collected works of Shakespeare digitized and downloaded as one big electronic text file. Or we might have a Web crawler take a snapshot of every web-page on the Internet. More often, however, investigators do not have direct access to the entire population. Also, it may not be necessary to study every datapoint to understand the features of the population; for example, not every grain of sand on a beach needs to be examined individually.

Instead, investigators take a **sample** and use the mean, variance, and shape of the **sample distribution** to make an unbiased estimate of the mean, variance, and shape of the underlying population from which it was sampled. Whether the population follows a normal distribution or not, the sample mean provides an estimate of the population mean. The variance of the sample distribution ($= SS/df = SS/(n-1)$) is always a bit larger than the population variance ($= SS/n$), reflecting the greater uncertainty involved in sampling.

So far we have mentioned two distributions: there is the population of interest, whose distribution is unknown, and the sample distribution, whose distribution is used to characterize the underlying population. The larger the sample, the better the sample will resemble the overall distribution—provided that the sample is taken in an unbiased manner, so as to reflect the composition of datapoints across the entire population.

Statistical Testing: Measuring the Variability Associated With Sampling

Comparing two different groups in an experiment is performed by asking whether the sample distribution of group 1 and the sample distribution of group 2 might plausibly have been sampled from the same underlying distribution. If so, then their apparent differences might have been due simply to the variability inherent in the sampling process. If not, then they appear to be "significantly" different from each other, within the bounds of the statistical test that is used. Most commonly in an experiment, groups 1 and 2 have similar variances and shapes, so we focus on asking whether the mean value of group 1 appears to be significantly different from the mean of group 2. Also, in most experiments, one group is considered the baseline or **control group** and the other is some experimental group that is being compared with it. So, in this discussion, we will assume that group 1 is the control group, acting as a reference group for comparison to other groups.

So, what we really asking is, first: **how much do sample means vary when samples are taken repeatedly from the same group, say group 1?**

A. DESIGNING YOUR EXPERIMENT

To assess this, we form a **third distribution**, i.e., the distribution of sample means for group 1. Whether the population resembles a normal distribution or not, the **Central Limit Theorem** says that as one samples repeatedly from the same population, as the number of samples grows toward infinity, **the distribution of sample means will approximate a normal distribution** (Fig. 3.4). The **distribution of sample means**, otherwise known as the **sampling distribution** (Fig. 3.4), provides an estimate of the population mean. Even without carrying out repeated sampling, if one has a quasinormal sample with adequate datapoints in it (>30 in most cases), the sample mean can be taken as a good estimate of the population mean.

The mean of the sampling distribution is simply the average of the sample means, and its standard deviation (most commonly called the **Standard Error of the Mean, or SEM**) is simply the sample SD divided by the square root of n, the number of datapoints in the sample (that is, SEM = (sample SD)$/\sqrt{n}$). Rather than the names "sampling distribution" or "distribution of sample means," I prefer the name **"skinny curve,"** because inside every sample distribution there is a skinnier curve that follows the normal distribution, which you will use for both power estimation and hypothesis testing. (The skinny curve is always narrower than the original sample distribution, since the standard deviation of the sample distribution is SD_s, whereas the skinny curve's standard deviation $= SD_s/\sqrt{n}$.) The 95% CI of the skinny curve comprises the region between two SEM below the mean and two SEM above the mean, and indicates that **there is a 95% chance that the true population mean resides within this range**. The **margin of error (MOE)** is usually defined as half the width of the 95% CI.

Statistical Testing: Comparing the Means of Two Groups Using the t-Test

We will discuss null hypothesis statistical testing and the t-test in detail in Chapter 9, but here is the gist of it: Suppose you have sampled heights from two groups, say, examined 1000 children living in large cities versus 1000 children living in rural areas. The mean of the city sample is 5 ft, and the mean of the rural sample is 5 ft 2 in. What are the chances that these two means are significantly different from each other? More precisely, what are the chances

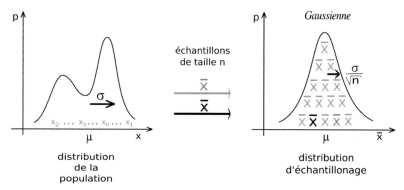

FIGURE 3.4   The Central Limit Theorem showing that repeated sampling of an arbitrary population creates a distribution of sample means x, also known as the sampling distribution or the "skinny curve," which approximates the normal distribution. Note that as the size n of each sample increases, the SEM will go toward zero. Also, note that σ here refers to the SD of each sample, not the SD of the underlying population. *Figure by Mathieu Rouaud, https://commons.wikimedia.org/w/index.php?curid=35724595 with permission.*

that the apparent difference that you have observed is simply due to sampling variation? After all, even if we had sampled twice from urban kids, the two sample means would not have been exactly the same.

To answer this question, conceptually, plot both the distribution of the urban sample and the rural sample, and construct the skinny curves for each one. Each skinny curve has a sample mean and 95% CI (Fig. 3.5). **Do either of the CI from one distribution overlap with the sample mean of the other distribution?** If they do *not* overlap (Fig. 3.5), there is less than a 5% chance that the differences are due simply to sampling variation, and we would say that the means are "significantly different."

Note that this testing framework does not assume that the population under study follows a normal distribution, but it does assume that the skinny curve is a reliable measure for constructing the CI. That means either that (1) the underlying population is normal or quasinormal, and at least 20–30 datapoints per group are sampled, or (2) the number of datapoints is large enough so that the Central Limit Theorem applies. If the population is highly skewed, as in Fig. 3.3, it is possible that hundreds or thousands of datapoints per sample will be required. A variety of statistical tests are often more appropriate than the t-test and will be described in Chapter 12.

OTHER IMPORTANT CURVES AND DISTRIBUTIONS

In contrast to the previous section that measures how the mean will vary across different samples, the chi-square distribution captures how the **variance** will vary for different samples taken from the same population. For any distribution of k independent datapoints, each drawn from the standard normal distribution, the SS (sum of squares) for the datapoints will follow the **chi-square distribution** with mean = k and var = 2k (Fig. 3.6).

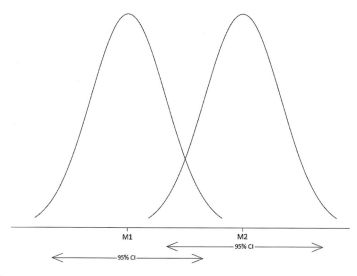

FIGURE 3.5 Shown are the sampling distributions (aka distributions of sample means, aka *skinny curves*) corresponding to two samples. The curves overlap, and even their 95% CI overlap. However, the mean of one curve does not overlap with the 95% CI of the other curve, so we say that the two distributions have significantly different means (see Chapter 9 for details). Note that this procedure is *not* carried out directly on the distributions of sample datapoints whose variability is given by SD_s, but rather involves constructing the distributions of the sample means, whose variability is given by $SD_s/\sqrt{n}$.

FIGURE 3.6 **The chi-square distribution.** *From https://commons.wikimedia.org/ wiki/File:Chi-square_pdf.svg with permission.*

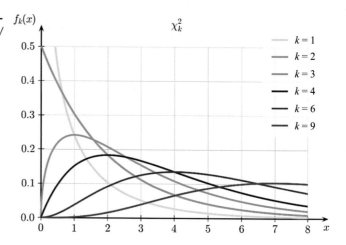

The chi-square distribution is also used to compare two distributions and decide if they are "significantly" different from each other. For example, to see if a particular distribution is approximately normal or not, one can plot the curve and superimpose a normal curve having the same mean and SD on it. The deviation of one curve from the other at each datapoint, squared, and summed up over all datapoints, is the SS. If the SS value falls beyond 2 SD above the mean of the chi-square distribution, then the normality test is said to fail.

The **F-distribution** describes the ratio of two chi-square distributions (Fig. 3.7). As we will see in Chapter 11, the F-distribution is used to compare the amount of variability observed BETWEEN experimental groups to the variability observed WITHIN each group, that is, the $SS_{between}/SS_{within}$ ratio. Note that the F-distribution describes the ratio of two chi-square distributions, so it is necessary to specify the df of the numerator as well as the denominator.

FIGURE 3.7 **The F-distribution.** *From https://commons.wikimedia.org/wiki/File:F_ distributionPDF.png with permission.*

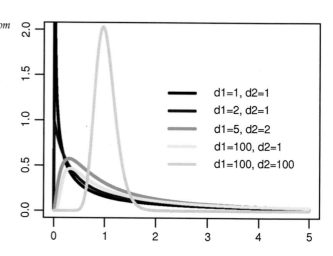

PROBABILITIES THAT INVOLVE DISCRETE COUNTS

This is the math of coin flips, lottery winnings, and so forth. For a coin flip, the probability of getting Heads = $1/2$, Tails = $1/2$, and p(H OR T) = $1/2 + 1/2 = 1$.

One can construct trees to see all possible outcomes of discrete probabilities:

H or T

HH or HT or TH or TT

HHH or HHT or HTH or HTT or THH or THT or TTH or TTT

For n independent events each with probability p of occurring, the expected number of events occurring follows the **binomial distribution** with mean = np and variance = np(1 − p). For example, if you buy a lottery ticket with probability of winning one in a million, then if you buy 10 tickets, the probability of winning is 10 in a million. As n goes to infinity, and as p goes toward zero, the binomial distribution will approximate a normal distribution.

The **hypergeometric distribution** describes the probability of k successes in n draws, **without replacement**, from a finite population of size N that contains exactly K successes, wherein each draw is either a success or a failure. (This is in contrast to the binomial distribution, which describes the probability of k successes in n independent draws, that is, **with replacement**.) For example, if there are 40 red and 60 green marbles in an urn, the hypergeometric distribution will allow you to calculate the probability of choosing 3 red marbles when you remove a total of 10 marbles. Note that the probability of choosing a red one is not constant, but changes over time depending on which marbles are chosen and as the size of the remaining population of marbles shrinks. The hypergeometric distribution is the basis for a class of statistical tests known as **exact tests**, which will be described in Chapter 12.

CONDITIONAL PROBABILITIES

Taking an HIV Test

Consider an HIV test that has 100% sensitivity, that is, if a person has HIV, there is a 100% chance that the test will be positive. The test has a small false positive rate, say 1%, meaning that if a person does not have HIV, the test will be positive anyway 1% of the time. You walk into a clinic and take the test. It is positive. What is the chance that you actually have HIV?

In equation form,

$$P(\textbf{Test} = \textbf{Pos} \text{ Given that } \textbf{Person} = \textbf{Pos}) = 1$$

$$P(\textbf{Test} = \textbf{Pos} \text{ Given that } \textbf{Person} = \textbf{Neg}) = 0.01$$

A. DESIGNING YOUR EXPERIMENT

We want to know

$$P(\textbf{Person} = \textbf{Pos} \text{ Given that } \textbf{Test} = \textbf{Pos})$$

As shorthand, P(A Given that B) will be written P(A|B).

The answer is…. that the question does not contain enough information to provide the answer! What we still need to know is the **a priori** or **prior probability** that the person being tested has HIV. A person who is in a very low-risk group and someone who is in a high-risk group will have very different risks of the test being a false positive.

The key is to use Bayes Theorem (which will be derived in Chapter 10; just accept it for now):

$$P(A|B) \times P(B) = P(B|A) \times P(A)$$

which is usually written as

$$P(A|B) = [P(B|A) \times P(A)]/P(B)$$

Note that P(B) is the sum of two nonoverlapping cases, namely when A occurs and when A does not occur, so P(B) can be substituted with the expression $[P(B|A) \times P(A)]$ $+[P(B|\text{not } A) \times P(\text{not } A)]$.

$$P(A|B) = [P(B|A) \times P(A)]/[P(B|A) \times P(A)] + [P(B|\text{not } A) \times P(\text{not } A)]$$

Substituting the terms of the HIV test, we now have:

P(Person = Pos|Test = Pos) =

P(Test = Pos|Person = Pos) × P(Person = Pos)/P(Test = Pos) =

P(Test = Pos|Person = Pos) × P(Person = Pos)/

[(P(Test = Pos|Person = Pos) × P(Person = Pos)) + (P(Test = Pos|Person = Neg) × P(Person = Neg))].

We know some of these values, but not all:

$$P(\text{Person} = \text{Pos}|\text{Test} = \text{Pos}) = (1) \times P(\text{Person} = \text{Pos})/[(1) \times P(\text{Person} = \text{Pos})$$
$$+ (0.01) \times P(\text{Person} = \text{Neg})]$$

Thus, we still need to know P(Person = Pos), the **a priori** or **prior probability** that the person being tested has HIV.

Suppose you lack any known risk factors—e.g., have never had sex nor received any blood transfusions. The prior probability of having HIV is very low, say, one in a million. Then

$$P(\text{Person} = \text{Pos}|\text{Test} = \text{Pos}) = (1) \times (0.000001)/[(1) \times (0.000001) + (0.01) \times (0.999999)]$$
$$= 0.000001/(0.000001 + 0.009999)$$
$$= 0.000001/(0.01) = \textbf{0.0001}.$$

A. DESIGNING YOUR EXPERIMENT

  The chances of being HIV positive are still only 1 in 10,000, even though the test was positive! If 10,000 similar low-risk individuals are tested, only 1 is expected to be a true positive, yet 100 individuals (= 1%) will receive a positive test result.

  Conversely, suppose you are in a high-risk group, and the prior probability of being infected is relatively high, say, 0.2 or 2 in 10. Then

$$P(\text{Person} = \text{Pos}|\text{Test} = \text{Pos}) = (1) \times (0.2)/[(1) \times (0.2) + (0.01) \times (0.8)]$$
$$= 0.2/(0.2 + 0.08)$$
$$= 0.2/(0.28) = \mathbf{0.71}.$$

  Thus, a positive test in a high-risk individual gives a much more confident result: 71% of such individuals will actually be HIV positive. We will give several more examples in Chapter 10 that show how conditional probabilities are essential for clear thinking about evidence.

ARE MOST PUBLISHED SCIENTIFIC FINDINGS FALSE?

  Conditional probability is also the basis for the famous critique written by John Ioannidis, "Why Most Published Research Findings are False," [1] which has been vastly influential in stimulating the current discussion about how to improve research reproducibility.

  To oversimplify his argument a bit, he considers that scientists in a given field explore a defined space of possible findings, which vary in their prior probabilities of being true. He constructs a toy model of scientific experiments, in which scientists "test" each finding by comparing an experimental group versus a control group, and asking whether the two groups are significantly different from each other with P-values of 0.05 or less. If so, the scientist publishes a paper stating that the finding is true. Note that the probability that an experiment will detect a true finding as significant (that is, its **power**; this concept will be discussed in detail in Chapter 6) is generally no more than 0.8 and often much less.

  Thus, the probability that a finding is actually true, given that the experiment gives a positive result, is:

$P(\text{Finding} = \text{True}|\text{Exp} = \text{Pos}) =$

$P(\text{Exp} = \text{Pos}|\text{Finding} = \text{True}) \times P(\text{Finding} = \text{True})/P(\text{Exp} = \text{Pos}) =$

$P(\text{Exp} = \text{Pos}|\text{Finding} = \text{True}) \times P(\text{Finding} = \text{True})/$

$[(P(\text{Exp} = \text{Pos}|\text{Finding} = \text{True}) \times P(\text{Finding} = \text{True})) + (P(\text{Exp} = \text{Pos}|\text{Finding} = \text{Neg}) \times P(\text{Finding} = \text{Neg}))].$

  Substituting in values for power = 0.8 and false-positive rate = 0.05, we get:

$$P(\text{Finding} = \text{True}|\text{Exp} = \text{Pos}) = 0.8 \times P(\text{Finding} = \text{True})/[(0.8 \times P(\text{Finding} = \text{True}))$$
$$+ (0.05 \times P(\text{Finding} = \text{Neg}))].$$

  How can we estimate the prior probability of a new scientific finding being true? Let us say that a finding that is reasonably surprising may have a prior probability somewhere between 0.1 and 0.01. This gives:

For prior probability of 0.1,

$$P(\text{Finding} = \text{True}|\text{Exp} = \text{Pos}) = 0.8 \times P(\text{Finding} = \text{True})/[(0.8 \times P(\text{Finding} = \text{True}))$$
$$+ (0.05 \times P(\text{Finding} = \text{Neg}))]$$
$$= 0.8 \times 0.1/[(0.8 \times 0.1) + (0.05 \times 0.9)]$$
$$= 0.08/(0.08 + 0.045)$$
$$= 0.08/0.125 = \mathbf{0.64}$$

Ioannidis argues that biases, low power, multiple testing, and other factors that are rampant in current experimental designs push the probability of a reported finding actually being true down further than these estimates (we will discuss these issues in detail in Chapters 8 and 10). Even for the a priori probability of 0.1, the presence of biases means that the vast majority of articles are reporting findings that have probabilities of being true <0.5, i.e., have less than a 50% chance of actually being true.

For prior probability of 0.01,

$$0.8 \times 0.01/[(0.8 \times 0.01) + (0.05 \times 0.99)] = 0.008/(0.008 + 0.0495)$$
$$= 0.008/0.0575 = \mathbf{0.139}$$

The estimated probability that a positive finding is true, for findings that have a prior probability of 0.01, is only 13.9% in the best-case scenario! This is a highly original (and somewhat controversial) way of looking at scientific experimentation, but it agrees with the empirical evidence that we reviewed in Chapter 1 and dramatically emphasizes the need for scientists to improve reproducibility of their published findings.

Reference

[1] Ioannidis JP. Why most published research findings are false. PLoS Med August 2005;2(8):e124.

Experimental Design: Measures, Validity, Sampling, Bias, Randomization, Power

MEASURES

Science is based on measurement. When the objects under study cannot be measured objectively, accurately, or precisely, we do not speak of science but rather enter the territory of other human endeavors such as arts, humanities, and spirituality. Measures can be described according to how different samples or subjects are compared with each other:

Nominal or **categorical** measures are those that simply assign labels or classifications; for example, we may assign tags of "male" versus "female" or "American" versus "Chinese."

Ordinal measures line up samples from smallest to largest. For example, one might say that a particular subject is taller than 99% of the others in the population under study. Another ordinal measure is to rate movies on a scale of 1−5 stars. Ordinal measures are not necessarily linear or uniform: A person who is taller than 40% of the population is not necessarily twice as tall as someone who is taller than 20% of the population. The difference in heights between two subjects that lie 20% apart (say, at the 40% point vs. the 20% point) is not necessarily the same at other places in the distribution (e.g., at the 80% point vs. the 60% point).

Interval measures are those that maintain an equal-interval ordering. For example, temperatures measured on the Celsius or Fahrenheit scales are interval scales since the distance between 20 and 40°C is the same as the distance between 60 and 80°C. Note that these scales do not have true zeros (e.g., 0°C is not absolute zero).

Ratio measures do compare samples against an objective zero point; for example, temperature on the Kelvin scale is measured from the absolute zero point. Ratio measures preserve linearity and uniformity. An object at 40K is twice as warm as one at 20K, and the difference between objects at 40 versus 20K is the same as the difference between objects at 80 versus 60K.

Data Literacy
http://dx.doi.org/10.1016/B978-0-12-811306-6.00004-X

Measures can be characterized in terms of **accuracy** versus **precision**. A thermometer that is **accurate** will, on average, give the correct value. Who is to say what the correct value is? This is handled by having agreed-on calibration and standards. In the case of fundamental measures such as mass and time, national and international organizations maintain standards. (As we will discuss later, most complex or "**inferred**" measures lack external standards, which is a major problem in experimental science.) In contrast, a thermometer that is very **precise** will give values that are consistent down to thousandths of a degree. Of course, a very precise thermometer might systematically give the *wrong* value—for example, it might tend to give values that are consistently warmer than the ambient temperature.

Only What Gets Coded Gets Noticed

When choosing a particular measure for your experiment, you need to consider two different and sometimes opposing principles. On the one hand, you want to choose measures that are simple, inexpensive to score, and that suffice for your own purposes. On the other hand, ideally, you would like your raw data to be available for others to use in the future, perhaps for purposes that you cannot even envision at present. However, this may mean incurring extra effort and expense to score additional parameters.

If I am studying the response of humans to experimental drugs in a clinical trial, a simple assignment of subjects to "male" versus "female" is probably sufficient. In contrast, if I am running an endocrinology, psychiatry, or gender-reassignment clinic, I need to score gender quite carefully; across multiple dimensions, each of which lies on a spectrum. Self-identified gender may include, for example, male, female, male-to-female, and female-to-male. In fact, in 2016 the New York City Commission on Human Rights officially recognized 31 different self-reported gender identities! Outward appearance may differ from self-reported identity and may be male, female, androgynous, or hermaphrodite. Chromosomal identity may be male (XY) or female (XX) as well as many chromosomal variants that are associated with altered behavior and health status (e.g., XXY, XXXY, XO). Sexual preferences and sexual behavior may also differ along dimensions and spectrums of their own.

Different Ways of Measuring Can Lead to Different Interpretations of the Same Data

You are the marketing manager for a small independent ice cream manufacturer. You are thinking of launching a new brand of ice cream and are wondering whether to launch flavor A or flavor B. To see which is better, you give samples of ice cream to 100 people on the street and use nominal measurements (that is, "prefer A," "prefer B," or "no preference"; see Table 4.1).

TABLE 4.1 Ice Cream Preferences Measured in
Categorical Terms

Prefer A	Prefer B	No Preference
36%	9%	55%

Of those who have a preference, four times as many prefer A as prefer B. What would we conclude? From this, we might tend to conclude that people tend to favor A.

To make sure, we repeat the same experiment with the same subjects but using ordinal measurements (we ask them to rate the product on a 1–10 scale; see Table 4.2).

TABLE 4.2   Ice Cream Preferences Measured in Ordinal Terms

Flavor	Rating
A	5.5 ± 1.2
B	4.5 ± 1.2

Each flavor is rated on a 1–10 scale, 10 is the best; ratings shown are mean ± SD.

A has a better mean rating on average than B, but the difference is small and not even close to being statistically significant. From this, we might tend to conclude that A and B are pretty much alike. So go ahead with launching product A, then? Or maybe even launch both products?

But we do not want to make a mistake; after all, our job is on the line. So, we repeat the same experiment once again, with the same subjects, but now **calibrate** the measurements against an **external standard**. That is, we not only ask subjects to rate A and B, but also ask them to rate a known flavor, namely, Vanilla (see Table 4.3).

TABLE 4.3   As in Table 4.2, but Using Vanilla as a Reference Flavor

Flavor	Rating
A	5.5 ± 1.2
B	4.5 ± 1.2
Vanilla	8.0 ± 1.2

Oops! Both A and B are way worse than vanilla! Neither of the proposed products is really any good, is it? This does not mean that the flavors will necessarily fail in the marketplace (maybe they can be endorsed by the latest boy band or Nobel Prize winner). Nevertheless, the example shows that different ways of measuring the same endpoint can lead to totally different interpretations of the data, even when the findings themselves do not change at all.

Primary Versus Inferred Measures

Wavelengths, distances, and other **primary** measures can be agreed on and compared against external reference standards. However, most measures in biological and social sciences are actually **inferred** and are invented by individual experimenters or by consensus within a given scientific field. Inferred measures are conceptual entities rather than concrete things that exist in the real world. A hallmark of an inferred measure is that it could be measured potentially in many different ways, not all of which will correlate or agree with each other and no single one of which will encompass all of the aspects of the phenomenon to be measured.

For example, in economics, "wealth" and "productivity" are inferred measures. Wealth could potentially be measured as the amount of money in one's bank account, or one's net worth, or one's annual income, or the value of one's house and property; each of these choices are a **proxy** for the underlying inferred measure, i.e., wealth.

In genomics, the "gene" is an inferred concept. Even though DNA sequences do exist in nature, the exact region encompassed by a "gene" depends on one's definition or purposes. Some would define the gene as the DNA sequence that encodes a protein; others would say that a gene is any region that encodes an RNA transcript. Some would include the upstream and downstream regulatory sequences that control the transcripts as being part of the gene (whether or not these are contiguous with the protein-coding region).

In neuroscience, "pain" and "stress" are inferred measures. There are many different ways to assay "pain" in rats. For example, one can place a rat on a hot plate and count the time until it jumps. Alternatively, heat can be focused on the tail, and the time until the tail flicks or withdraws can be scored. Chronic pain can be scored by injecting irritants into the hindpaw and observing behaviors such as flinching, licking, and biting of the paw. Given that "pain" is a very complex construct that encompasses multiple dimensions and multiple neural mechanisms, it is no surprise that each assay is a **proxy** that only probes a small aspect of a rat's pain repertoire.

If you are carrying out experiments in an established field, you will generally have your choice of one or several widely employed, well-characterized, and well-standardized assays, which serve as proxies for each inferred measure that you are interested in. For example, if you are measuring protein concentrations in a sample, the Lowry method and the Bradford method are both popular. If you are measuring general intelligence in adults, the Stanford-Binet IQ test is standard. It is indeed a good thing to employ standardized assays, yet it is important to understand the details of how these work and not to use them blindly. For example:

1. Remember that each assay is merely a proxy that is only a partial, imperfect measure of what you are really interested in. The Lowry and Bradford assays give similar results in many cases, but they detect different things: the Lowry assay measures the binding of copper to proteins, whereas Bradford measures the binding of a particular blue dye (which does not bind equally to all proteins but primarily interacts with basic amino acids such as arginine, lysine, and histidine). They have different sensitivities and are affected differently by the presence of detergents. Similarly, the IQ test certainly does not measure all types of intelligence. You need to understand the nuts and bolts of each assay well enough to understand not only how to use them but also to decide which assay is the most appropriate for a given experiment. Often no single assay suffices, and an entire panel of assays is needed to get a broad overview of the underlying entity that you want to measure.
2. Carrying out a standardized assay may seem to be a routine procedure—and a commercial kit or apparatus may even have a printed instruction sheet to follow. However, make sure that you establish a reliable **baseline** and that you measure not

only the experimental samples or subjects but also calibrate the assay using **positive** and **negative controls** (see Chapter 5). The assay should be able to detect measurements along the full range that is encountered. **Floor effects** refer to assays that are insufficiently sensitive, which cannot detect differences among samples because both lie near or below the threshold of detection. Alternatively, **ceiling effects** occur when the assay reaches a maximal plateau value and cannot detect any higher levels. The difference between an assay's detection threshold and its maximum also creates a preset limit on how big a measured change in value between two samples can be. Suppose an assay has a threshold of 10 units and a maximal value of 500 units. Any two samples can only differ by a maximum ratio of $500/10 = 50$-fold (even if the true difference is 500-fold).

3. Moreover, even "standardized" assays are surprisingly affected by parameters that you might normally think of as inert and residing in the background (more of this in Chapter 5). For example, many procedures are carried out at room temperature, yet depending on the time of year and the geographical location of the laboratory, room temperature ($^{\circ}$F) may vary from the mid-50s to the mid-80s. This can significantly affect chemical reaction rates and organismal metabolic rates. Studies also often fail to consider the season of the year, day of the week, and the time of day, even though these may influence the subjects being studied.

4. Similarly, relatively slight differences in experimental protocols may entirely change what is being measured. Consider, for example, attempting to induce stress by giving a series of shocks to a rat's tail. The outcome is entirely different depending not only on the intensity and duration of shock but also on whether the shock can be anticipated or not by the rat, whether it can be escaped or not, how many sessions they receive, the time lapse between sessions, and so on.

Tip: Any time you alter or adapt an experimental protocol, even in a manner that seemingly should be irrelevant, you need to reestablish and recalibrate the assay carefully.

5. Finally, it is important to recognize that just because an assay is standardized, this does not mean that it is aligned naturally or robustly with the phenomenon under study. Recall the Bradford protein assay mentioned earlier, which measures the binding of a blue dye to basic amino acids. Since different proteins vary considerably in their content of basic amino acids, the Bradford assay is strongly biased and may not be very accurate for any given protein. (This is not a problem when the Bradford is used to measure tissue extracts, which contain hundreds of different diverse proteins.) Similarly, consider a social science study that measures the rate of movement of individuals from one job to another, as a proxy for measuring job dissatisfaction. Under situations and in job areas that are relatively stable, people may indeed move from one position to another in search of greater satisfaction. Yet in other situations, e.g., in areas affected by hurricanes or social unrest, job movement may reflect predominantly larger forces rather than individual preferences. One cannot simply choose a measure that works in one situation and apply it automatically to another, without assessing whether it is a good fit for the specific case under investigation.

Derived Measures

Primary measures can be calibrated against reference standards, and inferred measures are concepts that are measured using proxies. However, many measures studied in science are **derived**, that is, they are entities that are not defined in terms of responses per se but in terms of **changes** that occur relative to a baseline response. A classic example is the phenomenon of long-term potentiation (LTP) assayed in brain slices, which has been very extensively studied and is thought to represent a form of information storage that underlies learning and memory [1,2]. LTP is defined as the **difference** between the size of the electrical field potentials measured under baseline conditions (in, say, the CA1 subfield of the dorsal hippocampus) and the field potentials measured at different times after the neurons have been stimulated with a barrage (tetanus) of synaptic inputs (Fig. 4.1).

Derived measures, despite their popularity and sophistication, have a major conceptual drawback that they are inextricably tied to the experimental paradigm in which they are measured. For this reason, it is difficult to compare one study against another or one brain region against another. Furthermore, it is very difficult to interpret changes in parameters or drug treatments done within a single experiment [2–5]. Experimenters generally carry out mechanistic tests by altering one set of parameters and observing how this affects the response, in this case, the magnitude of LTP that is produced. However, changing the strength or patterning of the stimuli or the ionic composition of the bath might not simply

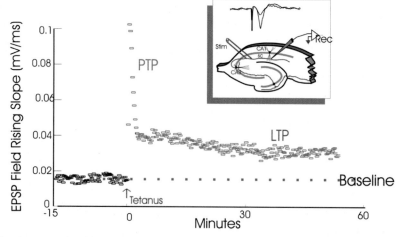

FIGURE 4.1   An example of long-term potentiation (LTP). The graph illustrates the analysis of a single field potential recording from a rat hippocampal slice. The inset shows the placement of the stimulating and recording electrode within the slice, and above, two raw traces of EPSP field potentials before and after tetanic stimulation. Field potentials were evoked by stimulation of Schaffer collaterals and recorded in stratum radiatum of the hippocampal slice (i.e., the Schaffer collateral/CA1 synapses). The individual points on the graph each represent the measurement of the rising slope of one EPSP field potential. *Black squares* indicate the measurements taken before tetanic stimulation. The *green squares* are measurements taken immediately after tetanic stimulation (PTP or post-tetanic potentiation), whereas the *blue squares* are measurements taken between 3 and 60 min after tetanization (LTP). Test stimuli were administered once every 30 s. Tetanic stimulation was the test stimulus given at 100 pulses/s for 1 s. Note how the amplitude of the EPSP field settles into a new, more elevated level after tetanic stimulation. *By Synaptidude, CC BY-SA 3.0, https://commons.wikimedia.org/w/index.php?curid=886855, figure legend modified.*

alter the amount of LTP that is observed—they actually might bypass or recruit new cellular and molecular pathways, and thus **alter the nature of the system under study**! Stimulating slices at 10 pulses/s versus 1 pulse/s can mean the difference between potentiating responses and suppressing them. Even stimulating slices at 10 pulses/s, at a steady rate versus given in periodic bursts, can alter not only the magnitude but also the mechanisms of LTP that is produced [5]. This limitation applies to any "black-box" approach to a system. We must abandon the naïve view that an experimenter can simply alter the inputs and observe the outputs and—by such means alone—can hope to surmise what the internal workings of the system must be.

Derived measures are very common in psychology and neuroscience. For example, memory is an inferred measure (studied using one or another standardized assay), whereas retrograde amnesia is a derived measure (since it is defined as a change from a baseline memory curve). Because derived measures are defined within the experimental setup used to measure it, it is very difficult to obtain robust or general results that can replicate across different conditions, assays, species, etc. Derived measures exist in other fields too, of course. For example, in economics, a "recession" refers to an acute decline in economic activity, such as a fall in Gross Domestic Product. But different recessions may not necessarily share the same underlying causes or mechanisms, e.g., those occurring during different stages of the business cycle in capitalist countries versus those occurring in different types of economies or governments.

How to make progress in studying derived measures? The answer is to tie the experiments to some observable phenomena, baseline, components, or responses that lie **outside** the experimental setup per se, yet are common to diverse situations in which the derived phenomenon may occur.

In the case of LTP, one might directly measure cellular and molecular events that are thought to be important in mediating both baseline responses and the induced changes—e.g., the number of AMPA receptors expressed on dendritic spines, the calcium concentration of the cytoplasm at dendritic branch points, the firing rates of interneurons synapsing on the neuronal cell bodies, or extracellular levels of neurotransmitters such as GABA or dopamine. The key is to understand how these events—which we might call **intermediate responses** or **candidate** mediators—are expressed, both during baseline conditions and as input stimuli and responses are altered. This helps to **open up the "black box,"** to examine some of the potential mediators, and to assess whether one LTP experimental setup is likely to be comparable to another. Clutch problems in one car versus another are more likely to be comparable if they both have manual transmissions!

VALIDITY

There are many different senses in which a measure may be valid. For example:

Face validity means taking the measure at face value—for example, a measure of anxiety that has face validity would be to ask a person if they feel anxious!

Interrater validity means that two different, independent scientists using the same measure on the same subject would obtain the same measurement. For example, two psychiatrists using the DSM-V manual to diagnose a patient with mental illness (Box 4.1) ought to agree on the diagnosis if there is high interrater validity.

BOX 4.1

THE STORY OF DSM-V

The *Diagnostic and Statistical Manual* is a checklist designed to assist psychiatrists in diagnosing mental illness. The first edition appeared in 1952; the fifth version, DSM-V, was released in 2013. The manual focuses almost exclusively on achieving high precision and reliability. For example, if a patient exhibits three or more symptoms (i.e., subjectively reported by the patient) from list A, which have lasted more than 6 months, and one or more signs (i.e., objectively reported by the family or observed behaviors) from list B, but do not have any of the exclusionary diagnoses in list C, then we say that the patient's diagnosis is X. The diagnoses are intended to have high face and interrater validity, but they are based entirely on observable surface features (Fig. 4.2). These diagnoses are not necessarily stable over time.

FIGURE 4.2   Patient: "I keep experiencing déjà vu." Doctor: "We discussed this yesterday." Psychiatrists make diagnoses based on observable signs and symptoms.

A. DESIGNING YOUR EXPERIMENT

BOX 4.1 (cont'd)

Nor do they probe or measure any underlying biochemical, genetic, imaging, electrophysiological, or other neurobiological parameters. The relationship between different disorders is based entirely on differences in their signs and symptoms. This is a problem since signs and symptoms are often shared across disorders (for example, depressed mood may occur in subjects suffering from depression, bipolar disorder, schizoaffective disorder, and posttraumatic stress disorder, among others).

The DSM-V does not provide a good starting point for the scientific study of mental illness, since a group of subjects diagnosed as "schizophrenic" may be very heterogeneous in terms of their underlying molecular or environmental causes. In fact, the National Institute of Mental Health has openly challenged and rejected the use of the DSM-V for research purposes [6]. Another problem is that the diagnoses in the DSM manuals are arrived at by consensus by a committee. The overall number of different diagnoses has mushroomed over the years, and it is not at all clear that the changes are related to better basic or clinical understanding of the diseases themselves.

Test–retest validity means that the same measure applied to the same subject at different times will obtain a stable reading. For example, if I administer an IQ test to an individual in March and May, the IQ should be roughly the same each time.

Split-half validity means if I split the sample randomly in half, both halves should have similar distributions of outcome measures.

Construct validity means that the measurement should be consistent with the known (or hypothesized) causes or mechanisms of the phenomenon. For example, if I attempt to measure intelligence by seeing how fast the subject can run a mile, this would have low construct validity, since most models of intelligence do not consider speed or physical fitness to be relevant.

Content validity means ensuring that the measures cover a range of aspects of the phenomenon; if I seek to measure intelligence, I probably should give subjects a panel of tests that cover a variety of different types of intelligence, not just (say) ability to solve jigsaw puzzles.

Concurrent validity asks whether a given measure gives results that agree, or are highly correlated with, other measures previously used.

Internal consistency means that different measures within a battery are consistent and do not conflict with each other.

Predictive validity is achieved if the measure is able to predict some related outcome in the future. For example, those who score higher on an IQ test might have a higher probability of subsequently being admitted to graduate school.

Generalizability occurs if the measure applies broadly and not just on a specific set or class of subjects. If I run a clinical trial to see if calcium supplements help prevent hip fracture and I validate a **biomarker** for hip fracture only on postmenopausal women in Chicago, the biomarker may be valid for that population but may not necessarily generalize to other populations such as men, younger adults, children, vegans, or rural Africans.

It is not always possible to choose a measure that satisfies all types of validity, particularly in fields that do not have well-developed theoretical models. If I am attempting to detect radio signals arising from outer space, I might make some educated guesses about the types of wavelengths and transmission patterns that might be sent (purposely or incidentally) by aliens. Alternatively, I might attempt to classify all signals as arising from one or another known source and seek to characterize signals that fail to fit into any conventional category. Either strategy is reasonable; neither is sure to work (even assuming that alien signals exist).

Errors and Variation

Closely related to validity, but even more concrete, are the concepts of **error** and **variation** that occur during the course of any experiment. Suppose you are measuring spontaneous activity in rats in their home cages using an automated camera. What kinds of errors can there be? Some errors are **systematic**: For example, the camera might not be centered but rather be directed to the extreme right side of the cage! Such a measure might be **precise**, that is, it gives a consistent exact number, yet it is systematically low (because it ignores activity on the left side of the cage) and so is not **accurate**. Alternatively, the camera might be out of focus or have low resolution, so that only the largest movements can be detected. In contrast, some errors are **random**. For example, power outages might cause temporary loss of data, or maybe the power source is not constant so it fluctuates unpredictably and makes the activity measurements fluctuate with it. These may result in data that get the right answer, on average, but with a lot of noise mixed in.

I remember that when I was in high school, my chemistry teacher came back from visiting a school in a third-world country and was amazed that their science class employed a yardstick that started from one (not zero). Thus, every measurement would be one inch too low!

Tip: When planning and analyzing your experiments, think about all of the possible sources of systematic and random errors that might affect your experiment—controlling them whenever possible, and looking at the data (to rule them out) when controlling them is not possible.

SAMPLING AND RANDOMIZATION

Suppose I want to learn whether blood pressure is related to diet. Out of a large potential population (all adults living in the United States), I might sample 1000 vegetarians versus 1000 meat eaters and observe whether the average systolic blood pressures of the vegetarians are similar or different from those of the meat eaters. Alternatively, if I want to take an interventional approach, I could choose subjects that appear to have similar features at the outset (say, 60-year-old males who live in Ohio and have equal **average** values of systolic blood pressure), then randomly assign one half to a vegetarian diet and the other half to a meat diet. Will the two groups become different in their average systolic blood pressure values at the end of the treatment period?

When sampling from a population, it is important to (1) choose an adequate number of samples and (2) ensure that the sample set captures the same distribution of attributes as it occurs in the underlying population. But we do not generally know the full number and range of attributes that vary in the underlying population, much less how these attributes are distributed or how they interact with each other! How to proceed?

One popular approach is to sample **randomly** across the entire population of interest. That is, each subject has an equal chance of being selected. We will discuss how to carry out randomization in the following paragraphs.

Another popular approach is to carry out **stratification** or **balancing** followed by random sampling. For example, if we are studying corn crops and suspect that elevation is a critical variable that might affect our results, we could divide the population of corn crops by farm elevation: 0–1000 ft above sea level, 1000–2000 ft, and so on, and then sample an equal number of plants randomly from each bin. That way, an equal number of samples will come from each elevation range, and each sample within a bin has an equal chance of being selected.

There are also more advanced strategies that are appropriate in some situations. For example, when picking items from a population, truly random sampling will result in occasionally picking the same item twice. This is often undesirable (one would not want to assign the same person to receive both active drug and placebo!) and so only distinct items are chosen (so-called **sampling without replacement**). Another advanced technique is **active learning**, in which not all items are equally likely to be chosen, but rather the sampling is biased in favor of those that are the most informative for the task at hand. Suppose I am trying to teach a person (or a computer) how to rate wooden boards in terms of their quality (acceptable vs. unacceptable). If I show them the best and worst examples, which are the easiest to judge, that will not help their skill as efficiently as showing them the cases in the middle, which are closest to the decision boundary.

Finally, in this era of Big Data, sometimes the entire population can be examined without the need to sample a subset at all! All public Internet Web pages and all Twitter comments that ever existed may be directly archived and studied with today's methods and computers. However, in most cases, random sampling is a much more efficient way to characterize a population unless you are specifically interested in the very rare or atypical cases.

How to Randomize

Errors in experiments are pervasive. There are the errors you know about, those you do not know about, and interactions among parameters, even parameters that you would never conceive of. For example, suppose you are studying an inbred mouse strain (Box 4.2) [7]. Different animal facilities, and even different rooms, may keep the mice rooms at different ambient levels of illumination. Does this affect your experiment? It might…. And certain strains of mice are affected more by illumination level than others! Randomization, if properly done, allows you to distribute variables (and errors) so that you do not make

BOX 4.2

PEOPLE STUDYING INBRED STRAINS OF MICE DO NOT NEED TO RANDOMIZE, DO THEY?

Because mouse studies are so common in biomedical science, let me single them out here. There is a prevalent belief among students that inbred strains of mice are genetically identical, and so they can be regarded as largely interchangeable for the sake of allocating mice to different treatment groups. Au contraire, inbred mice of the same supposed strain vary genetically across suppliers and within the same colony (because of, e.g., de novo mutations and copy number variants). Somatic mutations and transpositions happen during the lifetime of a mouse, so even cells and tissues within the same individual may exhibit genetic differences. Mice of the same strain, and even of the same litter, differ in many other ways as well, for example, prenatal position within womb, prenatal maternal events, size of family, number and type of siblings (and birth order), as well as the social structure of the family and colony. When examined in an experiment, females may be at any stage of the estrus cycle. Furthermore, things that happen to the mice before birth or during infancy can have very persistent, even permanent effects. For example, removing a preweaning pup from its mother can lead to permanent changes in DNA methylation of genes in the hippocampus [8]. If a mouse is handled by an unfamiliar or an inexperienced technician, it is likely to show a stress response, which may not occur when they encounter a familiar person or a familiar routine. If someone simply wheels a cage of rats into or even just past the mouse room, the sounds and odors of predators will have a strong effect on the mice for a long time to come! For all of these reasons, it is necessary to employ random sampling even on inbred mice.

Even random sampling of the mice that happen to be living in your animal colony at the moment does not go far enough to ensure robustness. For this purpose, it is desirable to carry out experiments on several different mice strains and stocks of the same strain, to repeat experiments using varied parameters (e.g., at different times of the year), and ideally even with the manipulations being done by different people and located in different laboratories. In short, robustness is built into an experimental design by making sure that similar results are obtained when one or a few parameters are deliberately changed.

Tip: If you are truly concerned with producing findings that hold up well over time and are generalizable, then randomization, and repeating experiments while varying one or a few parameters, should be built into the design.

systematic errors. As we will discuss later, randomization also allows you to make accurate estimates of statistical significance values.

To get statistically (and scientifically) reliable results, you must carry out true randomization and not **pseudorandomization**. An example of true randomization is assigning each subject a number using a random number generator, and then assigning all odd numbers to one group and all even numbers to the other group. Besides randomizing the choice of samples or subjects for study, more complex designs call for further randomization schemes. For example, it may be

A. DESIGNING YOUR EXPERIMENT

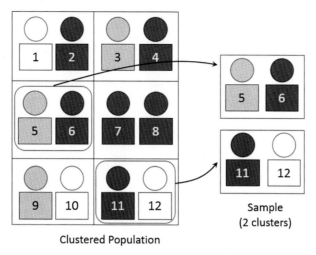

FIGURE 4.3 **Diagram illustrating cluster randomization, in which the clusters are chosen at random but not the individuals within clusters.** *By Dan Kernler (Own work) [CC BY-SA 4.0 (http://creativecommons.org/licenses/by-sa/4. 0)], via Wikimedia Commons.*

warranted to randomize the order in which interventions are given, or randomize which doctors will be assigned to the subjects within a given group.

In contrast, pseudorandomization is when the experimenter uses an arbitrary scheme, which appears to have no obvious relationship to the experiment. For example, I might run a clinic and assign all patients who come to my office on Monday, Wednesday, and Friday to one group, and those who arrive on Tuesday, Thursday, and Saturday to the other. I am assuming that there is nothing importantly different between patients who arrive on different days, but that assumption may not be true!

Cluster randomization is another type of pseudorandomization, often very convenient, but not necessarily free from bias (Fig. 4.3). For example, if I am studying students across various schools, I may choose schools randomly and compare certain schools against others (that is, I am studying students but only randomizing their schools). Another example is when experimenters choose litters of experimental animals at random and assign certain entire litters to the control group versus other litters to be experimentally manipulated.

And When NOT to Randomize

Sometimes true randomization can introduce problems. For example, suppose you are presenting a series of five shapes to a pigeon, in an experiment designed to measure how quickly a pigeon can learn the task. If the shapes are shown in true random order, once in a while there will be long runs of the same stimulus (say, seven circles shown in a row). Pigeons, like humans, are good pattern detectors and are likely to read in meaning where there is none (i.e., they can become superstitious gamblers just like humans). This will alter the behavior that you are trying to measure. To avoid this phenomenon, it is advisable to limit the length of runs of the same stimulus to no more than three or four, so the pigeon **responds** as if the run is truly random.

Another important limit to randomization occurs in social science experiments, where the very act of allocating interventions to different groups may alter the phenomenon you wish to observe. For example, suppose that in an observational study, it was noticed that future criminals were spanked more as children. Observational evidence is inconclusive because it is unclear whether children spanked more were misbehaving more. Suppose you try to resolve the issue by doing an experimental study: You randomly assign young children (age 6–9 years) to two groups: one will get a loss of TV watching for minor infractions and the other will receive a moderately severe spanking. You then test whether the children who were spanked will be more likely to be arrested for a crime during adolescence (age 16–18 years). What is wrong with this plan? Besides any ethical concerns (!), the spanking scheme requires the parent to alter their normal best judgment of parental care. This is likely to create more widespread changes in the relationship of parent to child, relative to what would occur under natural conditions when the parent, not a roll of the dice, decides the type of discipline.

A third limit on randomization (of treatments or order) is **hysteresis**. Treatments may produce changes in the samples or subjects that persist over time or may be frankly irreversible. If I am testing two different interventions sequentially on the same subject, there may be effects from the first intervention that linger and change the way that the subject reacts to the second intervention. For example, if I am testing mice on their ability to discriminate two odors, the learning should persist for days, weeks, or perhaps indefinitely. That may affect the rate at which the same mouse learns the same task (on two other odors) or even how they perform on different learning tasks. In drug trials, there is often a washout period between interventions to allow the effects of the first drug to fade, but without knowing all of the possible lingering consequences of the first drug, this may be difficult to ensure entirely. A **crossover design** ensures that the order of two sequential treatments is randomized; this is necessary but not always sufficient to balance for possible interactions between the effects of each treatment.

SOURCES OF BIAS IN EXPERIMENTS

Sampling bias arises when the sample datapoints are not truly representative of the population that they are intended to represent. This can occur because the sample is not obtained through proper random sampling procedures. It can also occur through **ascertainment bias**, which occurs when an entire subset of the population is missed or undersampled. For example, if you are conducting a poll to determine which candidate for president is favored, it makes a big difference how you reach out to people. If you approach people who are shopping at a mall or call people randomly on landline telephones, you will be favoring certain income, age, and demographic groups over others.

Unconscious sources of bias are generally dealt with through the use of **blinding**. Blinding means that the identity of a given sample or subject is unknown during the execution and analysis of the experiment. For example, if a patient is given a drug, generally clinical trials will do this in blind fashion, so that the patient does not know whether they are receiving an active drug, an experimental drug, or a **placebo** (e.g., a pill that looks and tastes like the drug but lacks active ingredients). The nurse giving the drug should also be unaware of what the patient is getting, lest they provide clues through their behavior. The physician who rates the clinical response of the patient should definitely not know whether the patient got the active drug or not!

And it may be less obvious, but **the person analyzing the data at the end of the study should also be kept in the dark about group identities**. It is all too easy to remove subjects or responses from the study as "anomalies" or "outliers" if they do not correspond to the expectations of the investigators. (The stories I could tell…sigh….) Even when studying test tube samples, animals, or surveys, the experimenters should be blinded to group identity until after the data have been fully analyzed, lest unconscious bias creep in.

Censoring bias arises when subjects drop out of the study or when some of the datapoints are less reliable than others. This can be a source of bias, since the way that you decide to handle these can affect or even dominate the overall results.

Finally, some unconscious types of bias cannot be dealt with through blinding or experimental design—at best, they can be recognized and acknowledged. These include personal and professional **conflicts of interests**. For example, you may be very reluctant to publish results that debunk your mentor's (or your) entire career. Commercial firms are particularly prone to constrain what their employees can and cannot publish; and even when the results are accurately portrayed, they tend to be "spun" and discussed in a way that favors the sponsor. Conversely, you may be very tempted to believe results that are positive or very striking, even if they are not fully validated, since these may lead to promotions, grants, and patents.

POWER ESTIMATION

How many observations (items, subjects, datapoints) need to be included in your study? You would like to be able to publish the study and have confidence in your results. This means having confidence that positive effects are real and are not simply produced by sampling error. This *also* means having confidence in negative effects as well. If you find that two groups are *not* significantly different from each other, are you sure that your experiment was sensitive enough to detect differences in the first place? **Power** refers to the probability that you will detect an effect in your experiment, provided that an effect (of a given prespecified size) does exist in your data.

It is not so easy to estimate in advance the sample size needed for an experiment. This is partly because you need to plug in estimates of various parameters and assumptions that you might not know accurately. Partly, it is because the entire exercise is designed to detect only one type of effect—that is, it assumes that you are looking for a difference in the mean value of one measurement, of a given size or larger, across two different treatment groups, and often the assumption is made that the data in each group will follow a normal distribution. Despite the limitations of power estimation, it is necessary to undertake this exercise at the time that you are planning your experiment. In fact, NIH and other funding agencies may require seeing your power estimation so that they feel confident the study will not be underpowered and hence fail to find an effect that is truly there. The technical aspects of how to do power estimation will be discussed in Chapter 6.

Given the strong constraints that most investigators face (time, effort, money, availability of specialized equipment), and given the difficulty in estimating the necessary sample size in advance, a popular approach is to undertake a small "**pilot study**" first, with the hope of expanding the study if the findings "look promising." This is not always a mistake, just like it is not ALWAYS a mistake to eat an entire bag of potato chips while promising yourself that you will diet tomorrow...but pilot studies arc almost always **underpowered** and too small to have confidence in the outcome, whether the outcome is positive or negative! Thus, if you follow up a pilot study that "looks promising," there is a high risk that you will be wasting your time. This is an important consideration for graduate students, who may spend a large fraction of their time carrying out pilot studies in search of observing effects that are "worth pursuing" in earnest. Pilot studies are like rehearsals, and not usually presented or even mentioned in the final publications, whereas "real" experiments are like concerts—they are out on stage for everyone to see. The incentive to carry out pilot studies is further exacerbated by well-meaning funding agencies who demand "preliminary data" in grant applications.

There *are* several valid, important reasons that one might carry out pilot studies:

1. To provide evidence that a new method can be carried out reliably in my laboratory. For example, if I am proposing to implant an electrode into the basolateral amygdala of a rat, I could perform surgery on a small number of animals to demonstrate my technique and see how long it takes them to recover fully.
2. To estimate the parameters that will be used in power estimation. For example, a pilot study of measurements carried out in control rats can provide estimates of the amount of inherent variation in values seen in the control group, as well as an estimate whether the data follow a normal distribution or are strongly skewed. These will improve the reasonableness of the power estimation used to decide how large each group will need to be in a "real" experiment.

Tip: Pilot studies do serve valuable purposes, BUT their small-scale outcomes should not be used as a way of deciding whether it is worthwhile to carry out a larger study.

References

[1] Bliss TV, Lomo T. Long-lasting potentiation of synaptic transmission in the dentate area of the anaesthetized rabbit following stimulation of the perforant path. J Physiol July 1973;232(2):331–56.
[2] Bliss TV, Collingridge GL. Expression of NMDA receptor-dependent LTP in the hippocampus: bridging the divide. Mol Brain January 22, 2013;6:5.
[3] Sanes JR, Lichtman JW. Can molecules explain long-term potentiation? Nat Neurosci July 1999;2(7):597–604.
[4] Lisman J, Lichtman JW, Sanes JR. LTP: perils and progress. Nat Rev Neurosci November 2003;4(11):926–9.
[5] Larson J, Munkácsy E. Theta-burst LTP. Brain Res September 24, 2015;1621:38–50.
[6] Insel TR. Director's blog: transforming diagnosis. April 29, 2013. http://www.nimh.nih.gov/about/director/2013/transforming-diagnosis.shtml.
[7] Bailey KR, Rustay NR, Crawley JN. Behavioral phenotyping of transgenic and knockout mice: practical concerns and potential pitfalls. ILAR J 2006;47(2):124–31.
[8] Turecki G, Meaney MJ. Effects of the social environment and stress on glucocorticoid receptor gene methylation: a systematic review. Biol Psychiatry January 15, 2016;79(2):87–96. http://dx.doi.org/10.1016/j.biopsych.2014.11.022.

When weighing an elephant, make sure it does not play with the scale!

5

Experimental Design: Design Strategies and Controls

"A FEELING FOR THE ORGANISM"

Being a scientist does not mean that he or she possesses any special knowledge, but that they follow a method. Ignorance is sometimes even a blessing in disguise, if it allows you to take a fresh perspective on old problems. On the other hand, when scientists are removed and detached from the phenomena that they are studying, their experiments can be formally correct yet their conclusions can be nonsensical (Box 5.1).

BOX 5.1

THE SCIENTIST AND THE FROG

There once was a scientist who studied frogs. One day, the scientist put the frog on the ground and told it to jump. The frog jumped four feet. So the scientist wrote in his notebook, "Frog with four feet, jumps four feet."

So the scientist cut off one of the frog's legs. The scientist told the frog to jump. Frog jumped three feet. So the scientist wrote in his notebook, "Frog with three feet, jumps three feet."

So the scientist cut off another leg. He told the frog to jump. The frog jumped two feet. So the scientist wrote in his notebook, "Frog with two feet, jumps two feet."

The scientist cut off one more leg. He told the frog to jump. Frog jumped one foot. So the scientist wrote in his notebook, "Frog with one foot, jumps one foot."

So the scientist cut off his last leg. He said, "Frog jump. Frog jump. FROG JUMP!"

So the scientist wrote in his notebook, "Frog with no feet, goes deaf."

(This is a very old joke, oft retold; this telling was taken from http://jokes.cc.com/funny-doctor/fj9nmq/the-scientist-and-the-frog.)

Data Literacy
http://dx.doi.org/10.1016/B978-0-12-811306-6.00005-1

As science becomes more standardized, the risk of losing "a feeling for the organism" [1] is greater than ever. Rats and mice are measured in defined quantitative assays using standardized apparatuses, often by experimenters who have never observed the normal behavior of these animals living in wild colonies. Molecular biologists manipulate molecules as samples of liquid pipetted in test tubes and may lack any feel for the molecules as physical objects with size, shape, mass, and interacting with other partners. If one is looking at clinical trial data, for example, you should have a clinician's understanding of what the patient population is like, how their diseases affect them, what the normal course of the disease is, how they received the treatments (and under what conditions), and so on. Otherwise, it is difficult to read the numbers that arise from the study, understand what dropouts and outliers might mean in the real world, or perceive the different factors likely to create variability in therapeutic response.

BUILDING AN EXPERIMENTAL SERIES IN LAYERS

For our purposes, the unit of scientific experimentation is the published article, which, more often than not, comprises a series of experiments that together make a coherent story. Let us consider experimental design strategies in successive layers.

First Layer

Let us follow scientist Jim Jones, who suspects that people eating a diet rich in fast food will be relatively poor in nutritional value, resulting in worse health indices than those eating a diet that avoids fast food altogether. He specifies the **population** of interest, say, adults living in Ohio, then carries out **random sampling** of (say) 10,000 residents and pulls out two **subgroups** for study, i.e., all of the ones who never eat at fast-food restaurants and all who eat fast food at least five times a week. Jim chooses some **outcome measures** to be **proxies** for health and nutritional status, say, weight, height, and systolic blood pressure. Each subgroup has its own distribution and mean value for each measurement. Statistical testing is carried out to see if the difference in mean heights is larger than would be expected to occur by chance (that is, if there was no true difference between the two groups, and the observed differences were due simply to sampling variability). If the probability of observing the difference in means by chance is less than 5%, the investigator concludes that the two samples ARE significantly different, at least in terms of heights. The same test is applied to weight and to blood pressure, and Jim proceeds to write up a paper reporting which, if any, measures differ as a function of fast food.

This experimental design is rather naïve and oversimplified. For example, how do we know that he obtained adequate numbers of people in each subgroup, to permit a reasonable comparison? If he does NOT see a significant difference, he might simply have studied too few subjects. That is, his experiment might lack adequate **power**. Differences in height are particularly likely to show small effects due to diet, which in turn would require more subjects per group than some measure that is likely to show big differences (like weight).

A. DESIGNING YOUR EXPERIMENT

Second Layer

Jim was smart to carry out random sampling of his overall population, but that does not mean that his two subgroups are necessarily comparable on all other factors that affect height, weight, or systolic blood pressure. For example, males may be more likely to eat a lot of fast food than females. So the high versus no fast-food subgroups may have a different proportion of males, not to mention different distributions of age, ethnicity, educational level, income level, and so on. These are potential **confounds** that may affect or even determine the outcome of the experiment. Since males weigh more than females (on average), a greater proportion of males in the high fast-food subgroup might cause the mean weight of that subgroup to be higher even if diet per se has no effect at all. Jim needs to identify and measure all confounds that he can think of, and **control** for them while analyzing the outcome measures. Almost every experiment is limited by the possibility that unknown confounds may exist, but at least, the known ones should be controlled as much as possible.

Another way to deal with the problems of sampling and confounds is through judicious choice of pairing subjects in experimental and control groups. For example, if we recruited a population of individuals and randomly assigned them to eat fast food or not, this **randomized controlled trial** would ensure that (assuming the population is large enough) the two groups will be balanced on most confounds. Alternatively, we could carry out an observational **cohort study** and identify a set of individuals who have similar distributions of age, gender, and so on and then follow their outcome measures over time (i.e., **prospectively**) as a function of their diet. Or, we could pair up individuals one-on-one with each other who have the same age, gender, ethnicity, and educational level but differ on diet, and then compare their outcome measures (this is called a **case-control study**). Or, we could carry out a **crossover study**, in which the same individuals agree to spend, say, 6 months eating one diet and then 6 months eating the other diet.

In theory, the best design is the randomized controlled trial with crossover interventions, in which subjects are chosen at random and then each person is assigned to eat one diet (in random order) followed by the other diet. However, the choice of design is generally limited by pragmatic considerations, such as cost, time involved, access to detailed information about the subjects, intrusiveness of the interventions, and so on.

Sometimes it is better to carry out a quick, inexpensive study than a 10-year behemoth that is designed "perfectly." For example, the type and prevalence of fast food eaten in the United States changes over a time course of years. The popularity of tacos from Taco Bell versus Whoppers from Burger King might vary considerably during the course of one's study. Among the same people eating fast food 5 nights/week, their diet might not be stable over time due to changes in restaurant popularity or changes in the composition of fast-food items. A study that takes many years to complete will be confounded by these changes. Studies of emerging diseases (such as Ebola) particularly need to be carried out rapidly, both because the setting of the disease may change rapidly and because society is waiting impatiently for the results of these studies to guide treatment and public policy decisions.

A. DESIGNING YOUR EXPERIMENT

Third Layer

Jim's experiment, if carefully sampled and analyzed, may well have internal **validity**, that is, his finding that one or more outcome measures are different between subgroups may be an accurate statement of the population that he sampled. But we really want findings that are also **robust** and **reproducible**, don't we! To ensure reproducibility, Jim ought to examine more than one sample of adults living in Ohio, to make sure that the two samples have the same baseline values in their outcome measures, and that similar **effect sizes** are observed across the two subgroups in each sample. To ensure robustness, there are a variety of things that Jim could do: He could employ different methods to carry out the sampling, for example, recruiting subjects on college campuses versus via online advertisements versus calling people at random versus going door to door. He could repeat the same outcome measures on the same individuals 6 months later, to see how stable the baselines are and how stable the differences are over time. He could employ different outcome measures that reflect nutritional status, for example, blood lipid levels. He could employ different experimental designs and see if the results are consistent regardless of the methods employed. Finally, he could sample additional populations, for example, those living in Texas, China, or Denmark.

Tip: In general, doing things more than once helps reproducibility; doing things in multiple ways helps robustness.

Fourth Layer

We would like an experiment not merely to produce findings but to shed insight into the underlying reasons and **mechanisms**. How might Jim adjust his experimental design to accomplish this? One simple way is to extend the range of relationships between diet and outcome measures. For example, he might divide up the population into more subgroups that vary along a continuum of fast food eating: no versus 1—2 versus 3—4 versus 5—6 versus 7—8 versus 9 or more fast-food meals per week. Perhaps nutritional status is only affected by extreme intake of fast food? Alternatively, perhaps the effects of fast food are clearly dose related and affect the majority of people in the population.

Examining **dose—response** curves is a great type of design in general, because it contributes to robustness and reproducibility AND gives hints regarding underlying mechanisms. Similarly, carrying out a **time course** of effects is another popular, powerful way to accomplish these goals. What are the effects of fast food after a week? A month? A year? Examining a range of time points provides far more insight than measuring just one time point.

But Jim could do even more to shed insight on mechanisms. He might measure levels of certain hormones that are known to be responsive to diet or that are known to modulate blood pressure. Finding such changes would not only strengthen acceptance of the basic findings but would also provide a starting point for further investigations of what is happening within the bodies of people eating fast food.

SPECIFIC DESIGN STRATEGIES

Paired Versus Unpaired Designs

Whenever you can reduce the amount of incidental variability *within* groups in an experiment, you improve your ability to detect differences that may exist *between* groups. One common way to reduce variability is to pair up subjects from experimental versus control groups. **Within-subject** designs are powerful because they make each individual serve as their own baseline control. For example, one can compare the same subject before versus after receiving drug A. Or, compare their response to drug A versus drug B. Or compare their response to drug A alone versus to drug A plus an add-on drug. **Case-control** studies pair up similar individuals from the experimental and control group. These are matched for demographic variables such as age and gender and possibly other variables relevant to the experiment. Animal studies often pair up animals from the same litter, i.e., **littermate controls**, who thus share the same parents and housing.

Even when the two groups A and B are not paired in terms of their selection, it is still advisable to pair up members of group A and B in terms of experimental protocols. For example, to minimize day-to-day variation in assays, equal numbers of samples in each group should be assayed or processed in parallel and in a random order. If you do not do this—say, you examine all the members of group A first then all group B—your design is susceptible to all sorts of possible time-varying confounds and incidental variables, which might interfere with your ability to detect true differences or cause differences to appear spuriously.

Tip: Pairing is usually, but not always, an improvement. When the items being paired do not all show a similar size and direction of effect, or when outliers are present, sometimes pairing can make significance values worse.

Stratification, Balancing, and Blocking Designs

An experiment should be representative of the population that you want to study. But how you achieve that depends on your population and on the question you are trying to answer.

Suppose you want to study the natural course of a disease that occurs predominantly in one gender, for example, autism ($5\times$ more common in males) or breast cancer ($100\times$ in females). Or suppose you want to study email behavior in congresspersons, who are predominantly male. If it is easy and inexpensive to study a sufficiently large number of subjects, the ideal would be to sample entirely at random from the entire population of interest. However, most investigators have finite research budgets. If the total number of subjects is limited, you may wind up with mostly one gender, with a smattering of the other—not enough to provide reliable findings with regard to the minority gender, yet enough to increase the heterogeneity within each group. This is the worst of both worlds!

Another option is to recruit equal numbers of males and females in your study. But this may not be optimal either, especially if (as in breast cancer) males may be rare and hard to recruit in adequate numbers. Male and female breast cancer patients differ in many respects, and unless your intent is specifically to compare males versus females, it may be better to restrict your study to females. This design allows the investigator to study twice as many females with the same budget, thus increasing the power to detect trends in the most common presentations of breast cancer.

More often, you may be interested in sampling the entire range of parameters present in the population but want to ensure that both common and less common values are adequately sampled. For example, say that you are studying the size of the average bet placed by visitors to a casino. If you have access to the entire data set of all visitors, you will be in a position to analyze the data according to any variables of interest—males versus females, young versus old, weekend versus weekday visitors, morning versus evening times, and so on. However, if you take a totally random sample of visitors, you may tend to underrepresent data related to weekdays (since most people attend on the weekend) as well as related to the extreme elderly (who represent a small proportion of visitors). To avoid this problem, you might utilize a **blocked design**, in which you place subjects into blocks (e.g., according to ages 21—30, 30—40, 40—50, 50—60, 60—70, 70—80, 80—90, 90+ years) and then randomly choose an equal number of subjects from each block. This strategy reduces the heterogeneity within each block and so has the strong benefit of improving the power of comparisons made within each block. For example, if I am trying to compare the overall size of bets in males versus females, my comparison is complicated by the fact that the males and females attending the casino may not have the same age distribution. If instead, I compare the males only against the females in the same age block, I will have greater ability to see any differences that are related to gender per se.

Even when you have not chosen a block design with regard to sampling and assembling your data set, you often can analyze your data according to blocks. For example, suppose you have run 10 separate trials of control mice versus mice that have been genetically altered in some manner and measured their response on a wheel running test. You could potentially pool all trial data together and make one big comparison. However, this introduces unnecessary heterogeneity due to the variable baselines and day-to-day changes in the assay. Better to compare the control versus altered mice in run 1, the control versus altered mice in run 2, and so on. Thus, the analysis of the experiment is **blocked by** experimental run. You can similarly analyze experimental data blocked by other experimental variables. For example, suppose you run out of a particular lot of an antibody or a commercial kit during the middle of an experiment. Even if the new lots have the identical catalog number, they may come from different lots (especially in the case of polyclonal antibodies), and it would be wise to analyze the results blocked by reagent—that is, compare the control and experimental groups from runs that share the same lot of reagent.

Factorial and Nested Designs

Factorial designs are appropriate when several different active interventions are being studied, which may interact with each other. For example, someone might study the effectiveness of a diet pill A versus placebo, a dietary regimen B versus usual diet, and an exercise regimen C versus usual exercise, on weight loss lasting at least a year in adult men with type II diabetes. A factorial design can be more efficient and informative than testing each effect individually. That means that each subject is randomly assigned to a specific **combination**

FIGURE 5.1 **Factorial designs for treatments A, B, and C permit examination of pairwise interactions between A and B, A and C, B and C, and a possible three-way ABC interaction**

of interventions ABC, and generally all combinations are studied. This allows the investigator to identify not only the **main effects** of treatments A, B, C, but also pairwise **interactions** between A and B, A and C, B and C, and a possible three-way ABC interaction (Fig. 5.1). Another example of a factorial design occurs when studying the effects of illumination, fertilizer, and temperature on plant growth. Instead of studying each variable separately, studying them in all combinations allows the investigator to learn the optimal pattern of growth (which may be different from the optimal levels found in each of the three variables studied in isolation).

Nested effects occur in situations when studying several variables A and B that are not independent. Suppose I am measuring the quality control of donated blood plasma. The plasma bags A are processed at a variety of blood centers B, so that As are nested within Bs. I will want to randomly sample plasma bags and to randomly sample from different centers; when it is time to analyze the data, the bags will need to be considered in the context of which center they come from.

Studying Large Versus Small Effects

The majority of published findings report effects that are modest at best in terms of **effect sizes**. They achieve statistical significance, surely, and may be credible, but think in terms of opportunity cost:

Tip: Time that you spend studying small effects is time that you are NOT spending studying large effects! All things equal, large effects are more likely to be robust and reproducible, more likely to be scientifically meaningful and important, easier to prove, and easier to analyze than small effects.

The first report of neurons responding to light patterns in the visual cortex involved only 24 animals and described the firing of neurons in qualitative terms, without presenting any numerical or statistical tests [2]; yet the receptive fields described by Hubel and Wiesel have been generally accepted and eventually led to their Nobel Prize. The earliest case reports of what would eventually become known as acquired immune deficiency syndrome (AIDS) involved only eight or nine patients, but the effects were striking: Rare diseases such as Kaposi's sarcoma and *Pneumocystis carinii* pneumonia, which were generally found only in elderly, chronically ill subjects, suddenly began appearing in young, previously healthy men. Again, no *P*-values needed to be calculated to conclude that something dramatic was going on. Large effects can also occur when there is a consistent pattern that is distributed across a large number of datapoints. For example, if you flip a coin 20 times and it comes up heads 20 times, this is a very big effect, even if no single flip is surprising when viewed on its own.

CONTROLS

As I am writing this section, my son just reposted an article on Facebook that accused Google of favoring Hillary Clinton and thereby attempting to influence the upcoming 2016 US election. The evidence is that a search on "Hillary Clinton" in Google brings up autocomplete suggestions that tend to route users to webpages that make positive statements about her. In contrast, the same search on Yahoo shows autocomplete suggestions that tend to route users to webpages that make negative statements. At first glance, the differences are striking and compelling. However, alternative hypotheses and crucial controls are lacking! One should type in "Donald Trump" in both Google and Yahoo and see if there is a reverse effect. It is possible that Google actively fosters positive sentiments or actively removes negative sentiments, for all candidates or indeed for any kind of searches. It is possible that Yahoo (and not Google) is manipulating the system. It is possible that the autocomplete suggestions are compiled from previous real user searches and that Google and Yahoo simply attract different types of users having different political views. The point is that apparently clear-cut effects can be shaky or outright false, when controls are lacking [3].

Most graduate students are taught the basics of positive and negative controls as part of their training, both in designing their own experiments and in critiquing the published studies of investigators in their field. Accordingly, this chapter will emphasize issues that, in my experience, are often overlooked.

Positive Controls

Positive controls are needed to check that instruments and measurement systems are calibrated and sensitive enough to detect true effects if they exist. They check that the system is working normally and check the assumptions of your experiment.

For example, let us consider a simple experiment related to long-term potentiation (LTP) in a hippocampal slice, asking whether adding a particular drug (say, a dopamine receptor agonist) to the bath will enhance LTP. A partial list of some positive controls may include the following:

1. Is the slice healthy? Does it exhibit baseline responses? Does it show LTP following electrical stimulation bursts?
2. Are other agents that are previously reported to improve LTP (e.g., forskolin) active in this system, for example, in other slices taken from the same animal?
3. Do we have confirmation that the drug is "good," i.e., dissolved fully and is stable, not degraded or oxidized?
4. Do we know that the drug is active in slices? Conceivably the drug might be active in whole animals because it is metabolically converted by passing through the liver but will be inactive when applied directly to slices.
5. Do we know that the tissue under study has the appropriate dopamine receptors and intracellular second messengers necessary to respond to the drug? Does the drug produce the expected biochemical changes in the test tube, in cultured cells, and in the slice preparation?
6. Does the drug produce a clear dose–response curve of effect? Does the halfmaximal effective dose correspond to the K_D of binding of the drug to the relevant dopamine receptor, as would occur if the effect is caused by the agonist binding to the dopamine receptor? (See Fig. 5.2 for an example of a binding curve.)

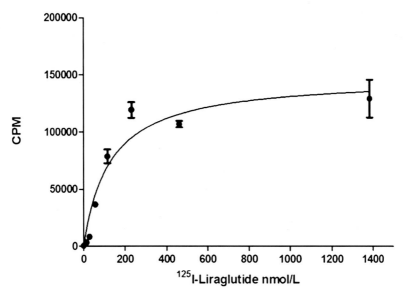

FIGURE 5.2 **Saturation curve of Liraglutide, an antidiabetes medication,** binding to its putative receptor, glucagon-like peptide-1 receptor. The equilibrium dissociation constant Kd for the binding 128.8 ± 30.4 nmol/L. Error bars represent 1 SD of the mean (N = 3). *Modified from Lv J, Pan Y, Li X, Cheng D, Liu S, Shi H, Zhang Y. The imaging of insulinomas using a radionuclide-labeled molecule of the GLP-1 analog liraglutide: a new application of liraglutide. PLoS One May 7, 2014;9(5):e96833. http://dx.doi.org/10.1371/journal.pone.0096833.*

A. DESIGNING YOUR EXPERIMENT

Even though this is a pretty simple experiment, there are many ways that the experiment could go wrong or give false results, unless appropriate positive controls are performed.

Note that the baseline excitability of the slice is a **background** variable, which is nonetheless critical to the success of the experiment. When a slice is initially prepared, it is electrically silent at first, then recovers and is stable for several hours, until the preparation eventually runs downhill. The experiment needs to be timed so that its time course falls within the window of health (as well as within the time course of LTP changes and the time course of effects produced by the drug). An animal's age, estrus phase, or postsurgical status may also be regarded as background variables when those parameters are not directly relevant to the study. In this sense, positive controls not only deal with the specific treatments and measurement calibrations but also help ensure that the background variables are in a state that permits stable measurements to be taken.

Assumption Controls

These are a type of positive control that is carried out to check and confirm whether the experiment is actually proceeding as expected. For example, take an experiment that seeks to test whether a diet rich in fish affects cholesterol levels. Assign two groups of Texans randomly: One group is told to eat their usual diet, and one is told to eat a filet of salmon or catfish daily. But it is possible that the "usual diet" group will be induced to think more about diet because they are participating in a study, so they might eat healthier in some way than they otherwise would. They might even increase their intake of fish! Conversely, the group told to eat more fish might balk at eating so many filets and wind up eating less than expected or even eating none at all. Furthermore, "usual diet" is expected to be the standard American diet, but diets vary enormously depending on where the study is done and with what populations. Thus, an **assumption control** would document the actual diets ingested in both groups, especially intake of fish, at baseline and during the course of the study, to see if the baselines and intended interventions were as expected.

Intersystem Controls

Intersystem controls are a great way to ensure robustness. They cross-check a finding in two or more different ways, using somewhat or totally different measures and methods (Box 5.2).

Negative Controls

One might think that negative controls are simply the inverse of positive controls. However, I find that they are much more difficult to design, for two reasons:
First, **a given treatment often has no unique or clearly best negative control.**
And second, **your negative control is often not negative!**
Suppose I am testing the effects of drug X on blood pressure in a mouse. The positive control is rather open ended—I can inject any known agent (or subject the animal to postures) that will raise and lower their blood pressure within my desired detection range. However,

---

BOX 5.2

AN EXAMPLE OF INTERSYSTEM CONTROLS

Suppose you are studying an enzyme, dicer, which cleaves double-stranded RNA, and want to study its expression in mouse or rat brain using a specific antidicer antibody and a specific method, immunocytochemistry. Not only should you carry out a series of positive controls to validate that antibody and that method, but also you should obtain complementary evidence using different antibodies and/or different methods. Some examples are as follows:

1. Does the antibody recognize a reference sample of purified dicer as well as dicer from brain extracts?

2. Do two different antidicer antibodies, raised against different epitopes (i.e., different sites on the dicer protein), recognize the same band on a Western blot?

3. If you immunoprecipitate with one antidicer antibody, will it pull down the band and be recognized by the other antibody on a Western blot?

4. Will the immunoprecipitate show dicer enzymatic activity (cleaving double-stranded RNA or pre-miRNA hairpins)?

---

what is my negative control? Do nothing and show that this does not affect the blood pressure? If drug X was dissolved in 0.1 mL saline and injected intraperitoneally (a fancy way of saying "into the belly") using a 23 gauge needle, then a better negative control would be to inject 0.1 mL saline and inject it intraperitoneally using the same type of needle. But it is quite likely that this injection will, in fact, alter the blood pressure of the mouse! It is likely to raise blood pressure transiently, because of the stress associated with handling the mouse. This might interfere with my ability to detect the effects of drug X, depending on the magnitude, direction, and persistence of the negative control injections. In such a situation, it would be better to change the experimental paradigm entirely. For example, one might use an indwelling catheter or osmotic pump to infuse the drug (vs. saline) slowly and steadily over a long period of time.

Many drugs are not soluble in saline and have to be given in DMSO or ethanol instead. The material used to dissolve an active agent is known as **solvent, adjuvant,** or **carrier**. But DMSO, ethanol, and even saline are active agents in their own right. Not only might they have direct effects on blood pressure, but they might also alter the physiology of the mouse (e.g., ethanol in sufficient doses can dilate skin blood vessels). Even if drug X does not interact directly with blood vessels, its action is likely to be qualitatively or quantitatively different in animals whose skin blood vessels are dilated.

Tip: One cannot ascertain the "specific" response to drug X by simply taking the response to drug X and subtracting the response observed in the negative controls! Instead, the experimental protocol needs to be adjusted and optimized to ensure that the negative controls are as inactive as possible.

A. DESIGNING YOUR EXPERIMENT

Some controls are both positive and negative at the same time. For example, investigators often measure the levels of a protein of interest in Western blots, with two or more treatment groups represented on adjacent lanes, and subsequently measure the levels of a reference protein such as GAPDH or beta-actin in the same lanes. The latter proteins act as both negative and positive controls. They are negative insofar as they are not expected to change with experimental treatment but positive insofar as they confirm that all lanes received equal loading of cell extract.

Sometimes the best choice of negative control is not at all obvious at the time that an experiment is planned. Some years ago, I was carrying out pilot studies to understand the toxicity in neurons of beta-amyloid peptide 1–42 (BAP), a peptide that accumulates in the brain of patients with Alzheimer disease. We introduced different doses of BAP into the medium surrounding cultured rat cortical pyramidal neurons and measured the extent of cell death 24 h later. But what should the suitable negative control be?

One negative control would be to add nothing and observe the baseline rate of cell death in the cultured neurons. (This ensures the health of the culture, considered as a background variable.) A slightly more specific negative control would be to add equal amounts of some inert protein (say, bovine serum albumin) to control for the amount of protein in the medium. More specific still would be to add a peptide that consists of the same 42 amino acids found in BAP but expressed in scrambled order. Another possible negative control might compare the effects of BAP with those of a different amyloid peptide, say, intestinal amyloid peptide (IAP), but we had no scientific reason for expecting that IAP should necessarily be negative! Alternatively, one might test a series of BAP peptides but mutated to have altered sequences, in which individual amino acid residues are altered (say, in one mutation, an alanine at position 20 is altered to a tyrosine). This manipulation is even more specific and, based on structure–function studies in other proteins, might be a promising way to shed light on biological mechanisms. However, at the time, we were not sure in advance which (if any) mutations will necessarily alter toxicity, or even if the primary peptide sequence is important for toxicity (as opposed to the peptide's conformation or aggregation status). These examples illustrate that choosing appropriate, relevant, and informative negative controls is often not an easy task.

SPECIFIC, NONSPECIFIC, AND BACKGROUND EFFECTS

We have pointed out that it is helpful to identify the possible sources of variability in the experimental setting and its outcome. One way to do this is to divide the phenomena that occur during an experiment into three classes: **Specific effects** are the responses that occur in the experimental group(s), whereas **nonspecific effects** are those responses that occur in the negative control group(s). **Background effects** are baseline responses that occur prior to, or in the absence of, any active intervention, as well as other variables that support the setting but are not considered to be an active participant.

In the case discussed above, of testing the effects of a dopamine receptor agonist drug on LTP in hippocampal slices, the **specific effects** are the measured outcomes seen when the drug is applied in the slice (i.e., the amount of LTP produced, plus any other variables that are measured by the investigator). The **nonspecific effects** are those seen in the negative control group (e.g., the amount of LTP elicited in slices given saline instead of active drug). The **background effects** include not only the baseline excitability of the slice (i.e., the size of the evoked potential when given a single stimulus) but also the volume of the bath that it is submerged in, the temperature and ionic composition of the buffer, the oxygen content of the chamber, and so on.

The specific effects, i.e., measured outcomes observed in the experimental group, do not occur in isolation but rather occur *on top of* outcomes that are observed in the negative control group. Recall that when injecting a drug into an animal, you produce not only specific effects on blood pressure but also nonspecific effects as well (due to the volume and composition of the solvent), and that potentially the specific effects will be affected by the nonspecific effects, as well as the baseline effects (e.g., the resting blood pressure at the time of injection). Again, I emphasize that you cannot reliably ascertain the effects of the drug simply by subtracting the measured outcome in the control group from the measured outcome in the experimental group.

And this applies generally, which is what makes even the simplest experiment potentially so complicated. **Specific, nonspecific, and background effects not only cooccur during an experiment, but they also all potentially influence the outcome, and they all potentially interact with each other! And the outcomes can be influenced not only by the individual effects themselves, but also according to how they interact** (Boxes 5.3 and 5.4). This is a major reason that so many experiments lack robustness and reproducibility. Investigators need

---

BOX 5.3

TRANSFECTING GENES IN CULTURED CELLS

A common type of experiment in biology involves introducing DNA into cultured cells so that the DNA is replicated within the cells and produces RNAs whose sequence is encoded by that DNA. I will try to give an example, using the minimum amount of jargon, to focus on the main points related to experimental design. Certain RNAs, called microRNAs, bind to certain other RNAs, called mRNAs and decrease their abundance. Let us say I want to test whether a given microRNA sequence, say mir-16, can effectively inhibit the level of a particular mRNA, say p53, within the environment of a living cell. One popular way to test this is to introduce (or transfect) a circular piece of DNA

(called a plasmid) that encodes mir-16 into cultured cells growing on a dish, together with another plasmid that encodes p53. If done correctly, the cell should express both the microRNA and the mRNA that has the expected target site. The goal of the experiment is to test whether the presence of mir-16 will be associated with a decrease in the level of the target mRNA, relative to some negative control baseline condition (for example, relative to the case where the mRNA is expressed but no mir-16 was introduced).

Recall that this experiment is based on a **derived measure** (Chapter 4). That is, the primary outcome is the *difference* in target mRNA levels between the group transfected

Continued

BOX 5.3   *(cont'd)*

---

with miRNA (a specific effect) versus the levels observed in the negative control group (a nonspecific effect).

As positive controls, I need to check that the transfection procedure did introduce the plasmids into the cell efficiently; that the transfection did result in production of mir-16 (and ideally I should estimate roughly how much is produced); and that the transfection also resulted in the formation of detectable p53. Despite the word "nonspecific," the nonspecific effects are actually sophisticated responses of the cells to being manipulated that involve important biological and biochemical pathways. **Some of these responses stimulated in the negative control may also be involved in mediating or modulating the "specific" effects being measured!**

1. For example, to induce cells to take up the plasmids, a popular method is to mix the plasmids with lipid mixtures; another is to mix them with calcium phosphate precipitates; and these complexes are then placed in the bath surrounding the cells, in both experimental and negative control groups.
2. Often, serum is normally part of the bath as a nutrient but is removed during transfection, which causes the cells in both groups to exhibit a stress or starvation response.
3. Once introduced into the cell, the plasmids are detected by surveillance proteins that detect the presence of foreign DNA and can elicit a cell stress response that shuts down production of most proteins in the cell or even lead to cell death.

Thus, nonspecific effects are both nontrivial and noninert in transfection

experiments. How to deal with this? One strategy is to measure the level of p53 only after waiting for several days, hoping that the cells have largely regained their equilibrium following the "insult" of transfection. Another is to compare transfection of mir-16 to negative controls that are matched as closely as possible to the experimental group. For example, transfecting mir-16 versus a mutated mir-16 (differing in only one nucleotide) is likely to share most, if not all nonspecific effects (you, of course, need to check that the change in one nucleotide does affect the affinity of binding of mir-16 to p53).

The background effects are also critical in transfection experiments:

1. For example, the experiment is done in some cell type chosen by the experimenter, who often picks some "standard" cell type (e.g., HEK293 or HeLa) that grows quickly and transfects well. But it is far from irrelevant which cell type is chosen, since each cell type expresses hundreds of miRNAs and thousands of mRNAs whose levels vary widely from one cell type to another. If the investigator chooses a cell type that already expresses mir-16 at high levels, transfecting more mir-16 may result in relatively little extra production and hence may appear to cause little suppression of p53 relative to negative controls. Similarly, if the cells already express a lot of mRNAs that bind mir-16, they may act as sponges and "soak up" the available mir-16, thus preventing mir-16 from having an effect on the transfected p53.
2. Different cell types also vary in the way they process and respond to foreign

---

BOX 5.3 *(cont'd)*

DNA, and different cell types are likely to produce different amounts of the plasmids.

3. Finally, transfecting certain specific microRNAs into certain cell types may cause drastic changes, for example, the cells may be induced to die or differentiate into neural cells.

I hope it is evident from this example that even quite simple, common, routine experiments can create a bewildering array of effects that need to be optimized and controlled carefully to interpret the findings in any meaningful way. Besides the "specific" effects of interest, the system will exhibit "nonspecific" and "background" effects that often interact in a very complex manner with the mechanisms that mediate the "specific" effects under study. The measured outcome (i.e., the percent inhibition of p53 by mir-16) will vary depending on the exact cell type used in the test, the exact transfection method, the exact dose of plasmids, and the exact time point of measurement, among other factors.

Tip: Do not think that any experiment can just perturb ONE variable and measure ONE effect!

BOX 5.4

THE PLACEBO EFFECT AS A NEGATIVE CONTROL

In clinical trials, sometimes patients are assigned to treatment A versus no treatment at all, but generally it is better to use an inactive treatment (**placebo**) as a negative control, so that patients do not know whether they are receiving the experimental treatment or not. A sugar pill is the classic placebo, although if the active drug has a distinct taste or effect (e.g., may make the heart race or discolor the urine), the placebo may be chosen that has similar effects. Often patients tend to get somewhat better in the placebo group relative to no treatment, although sometimes they get worse, in which case the placebo may be referred to as a "**nocebo**."

After years of neglect by the medical community, there is now increasing interest in understanding the mechanisms that underlie the placebo effect. This interest has been further stimulated by the frustrating experiences of pharmaceutical companies that have been testing new antidepressant drugs and find that they often produce significant improvement in symptoms—but so do the placebos! And not infrequently, the net difference between drug and placebo is not statistically significant. What does this mean? Does it mean that the drugs they are testing are inactive?

To date, most discussions of the placebo effect tend to discuss it in psychological terms, that is, in terms of the subjects' beliefs and expectations. However, I would like to look at placebos on the same footing as any other

Continued

A. DESIGNING YOUR EXPERIMENT

BOX 5.4   *(cont'd)*

negative control: in terms of interacting specific, nonspecific, and background effects that any treatment produces in any experiment:

1. The **specific effect** of an experimental drug is its biochemical effects in the body. Often the specific short-term and long-term effects of a drug are not entirely clear. For example, antidepressants seem to share the property of affecting monoamines in the short term, but patients often require weeks to show clinical improvement, so presumably antidepressants cause long-term effect(s) that are poorly understood, and that may not necessarily be directly related to the immediate changes in monoamines. One or more of these long-term effects may (together or separately) affect depressive symptoms.

2. **Nonspecific effects** are those associated with patients in both placebo and active drug groups. These may be related to conscious beliefs, but they also include the physiological consequences of being placed in a therapeutic environment. Is the nurse nice and empathetic? Is the clinic a welcoming place? Does the environment decrease, or increase, the level of arousal, stress, and trust produced in the subjects? The body has autonomic mechanisms—the fight-or-flight sympathetic system is associated with high cortisol levels, whereas there is also a prosocial trust system associated with oxytocin.

3. **Background effects** include the fact that the body has a sophisticated immune system, inflammatory responses, and central nervous system responses that normally allow people to heal

themselves and regulate normal moods even without any treatment at all. Drugs rarely cure anything outright but rather modulate the body's intrinsic systems. An antibiotic rarely kills every single bacterium within the body; rather, it generally slows down the load of bacteria so that the body's own macrophages and other immune cells can remove the rest. A cognitive enhancer might accelerate learning, but the drug does not learn anything on its own, it merely modulates a process happening in the brain.

4. Clearly, the efficacy of a placebo depends on background effects—the patient's brain must have its normal mood-regulating mechanisms to avoid depression. And it depends on nonspecific effects too: High stress (high cortisol levels) inhibits the immune system, suggesting that the mere removal of stress should facilitate healing. So anything in the "nonspecific" therapeutic environment may derepress both "specific" and "background" effects on healing. Importantly, **it is possible that the biochemical mechanisms that underlie placebo response may be partially or even largely overlapping with the "specific effect" mechanisms that are triggered by antidepressant drugs**. This is yet another situation in which one cannot simply subtract the placebo response from the drug response to ascertain the "net" efficacy.

From this perspective, clinical trials might be better designed to maximize the overall effect of the drug being studied than to minimize the placebo effect. However, such a

---

BOX 5.4 *(cont'd)*

---

strategy would require investigators (and regulatory agencies) to cease requiring that the drug be demonstrated to be statistically more effective than placebo. An alternative, politically safer route would be to focus on a different goal altogether, such as identifying the subset of depressed patients who are most likely to show the greatest benefit from the drug.

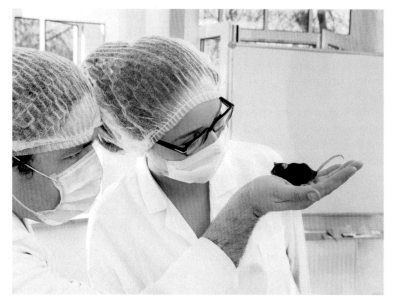

FIGURE 5.3   "Mickey always liked you best!".

to choose their negative controls judiciously and pay attention to background factors, even those that are easily overlooked and taken for granted, such as the gender of the person handling the experimental animals [5] (Fig. 5.3).

SIMPLE VERSUS COMPLEX EXPERIMENTAL DESIGNS

Scientific articles published in high impact journals such as *Science* and *Nature* tend to have extremely complex experimental designs and represent many man-years of effort by a dedicated team of experts. It is hard for me to argue against this path. However, if your goal is to perform experiments that are **reproducible** and **robust**, I strongly favor simpler designs whenever possible.

For example, a recent pair of articles in *Nature* were published back to back in the same issue, authored by two laboratories, which used extremely complex (and essentially identical) protocols to manipulate the activity of particular neurons in a restricted brain region, but when the mice were given different behavioral assays intended to model depressive-like behavior, each group reached opposite conclusions about the role of particular neurons in terms of whether they foster versus prevent the behaviors [6,7]. So even within the same experimental setup, the conclusions were not robust. This problem exists even with simple designs, of course, but is far worse with complex designs. Every aspect, step, or parameter of an experiment generates or contributes to three types of effects: specific effects, nonspecific effects, and background effects. Each step of an experiment needs positive and negative controls. Furthermore, each of these effects interacts with all of the others, and these interactions, too, require controls!

HOW MANY TIMES SHOULD ONE REPEAT AN EXPERIMENT BEFORE PUBLISHING?

Power estimation suggests how many datapoints should minimally be tested in the experimental and control groups. But how many times should one repeat an entire experiment before publishing?

There is no single answer or practice that everyone follows. A famous neuroscientist, Jerome Lettvin, once told me that he believed in carrying out preliminary experiments, or "rehearsals," to optimize the conditions, then performing a single definitive experiment or "concert performance" that would be reported in a paper. This is not the norm, however. Most investigators will carry out at least two experimental runs done the same way, to ensure that they are able to repeat at least the gist of the main results.

Tip: When repeating an experiment, double the groups: rerun the experiment using the exact same parameters, but also change one parameter in a second set of groups run in parallel.

That is, conduct the experiment both ways in the second run, to see if the results depend critically on the choice of this parameter. For example, if the original run tested a drug applied at 10 mM that is dissolved in 1 mL of saline, one might repeat the experiment using runs that compare different doses (e.g., 5 vs. 10 mM) or dissolved in different volumes (0.5 vs.1 mL). This balances reproducibility and robustness, and if the outcome changes, clues you in to which parameters may be important.

SOME COMMON PITFALLS TO AVOID

Caveat: Primary Versus Secondary Findings in Scientific Papers

People talk of findings in publications as if they were all equally important. However, generally, a scientific article will have one or a few primary (major) findings. In addition, the article will also present additional follow-up experiments or analyses that point toward possible explanations of the data or future directions for investigation. These secondary findings are often presented at the end of the results section or even in the discussion section. They often lack the careful attention to positive and negative controls, and optimization of all parameters, that was devoted to the primary findings. It should be no surprise that secondary findings are especially likely to be nonreproducible!

Caveat: Commercial Proprietary Kits and Reagents

When I was a PhD student, if someone wanted to use an enzyme in their experiment, they had to purify and characterize it themselves. It might take 2 years just to make enough for an experiment! Today, companies readily provide reagents and kits for nearly any common biomedical situation. This is a great boon, but … you do not personally know who made them, or how, or how they checked their results. Look at the guys and gals down the hall from you—would you trust them to make the critically important reagents for your thesis project? Believe it or not, some companies sell new products, even before they are fully tested or validated. This happens with antibodies all the time—they might be tested in one assay but then sold widely and used in conditions that are not appropriate. Believe it or not, many commercial products are plain crap, even those that are widely used and cited in publications. Furthermore, many reagents are proprietary, meaning that their composition is a trade secret. So no one can reproduce your results unless they use that exact reagent—and manufacturers often modify their products without notice. It is impossible to make sure that there are no errors or side effects if you do not know what is in the reagent. At a minimum, when reporting a commercial reagent in a publication, you should report the product number, and preferably the lot number too (see Chapter 14).

WHAT TO DO WHEN THE UNEXPECTED HAPPENS DURING AN EXPERIMENT?

This happens all the time! It could be a spurious finding; it could be real but uninteresting; or it could be interesting, yet a distraction from the present context. I cannot tell you what to do when you encounter an unexpected or incidental finding (Fig. 5.4), nor how to predict which are worth following up. However, I CAN tell you that without having full and proper controls in your experimental design, you will not be able to take them seriously and assess them properly when they do happen. For example, when my group was characterizing the expression of dicer protein in dendritic spines in the brain [8], we noticed that there was also immunoreactivity in the nucleus in some types of neurons. Although we did not pursue that observation further, we did describe the finding in the paper in some detail, to provide the basis for others to follow it up. One person's incidental finding may acquire new significance in the light of other findings made in the future and may lead to another's main research program.

SHOULD EXPERIMENTAL DESIGN BE CENTERED AROUND THE NULL HYPOTHESIS?

One of the most pervasive myths about experimental design is that the key task of the scientist is to pose a **null hypothesis** and to test this via **statistical significance testing**. In the case of Jim Jones' diet study, the null hypothesis is that the subgroups (eating no vs. high fast-food diets) will *not* differ in their mean weights, heights, or systolic blood pressure readings. According to the myth, if the null hypothesis is rejected, then Jim is entitled to conclude (1) that the subgroups *do* differ on one or more of these outcomes and (2) that the finding

**"Of course, we could just go to lunch ...
and forget we ever saw
this disturbing market research."**

FIGURE 5.4   One way to handle an unexpected finding.

provides support for his scientific hypothesis (i.e., that people eating a lot of fast food will have relatively poor nutritional status).

Why is this a myth and wrong or at least misleading? Let me count the ways!

First, remember what we said in Chapter 2: **A scientific paper should be like a court trial**—persuading the jury of your peers beyond a reasonable doubt that your findings cannot be explained by any alternatives that you (or they) can think of. A typical experiment comprises several different alternative ways and methods of asking the same question, different doses, time points, etc. These add up to a whole that lies beyond any single statistical test.

Second, statistics, hypothesis testing, P-values, and so on allow you to estimate one thing and one thing only—whether any observed differences between groups can be reasonably explained by sampling variability. That is, had you sampled the no fast food group a second time, what would the chances have been that you would have observed the same value as the mean of the high fast food group?

Tip: Rejecting the null hypothesis makes you reasonably sure that your effects are not likely to be simply due to sampling. It does not rule out any other explanations for the

observed differences, such as confounds, systematic errors, random errors such as equipment failure, or the possibility that your results are skewed due to outliers.

Third, statistical tests do not establish the truth or persuasiveness of a scientific hypothesis! A scientist wants to know the likelihood of a given hypothesis being true, given the data that are observed (= P(my Hyp | observed Data)), but statistical testing gives us something quite different, i.e., the probability of observing those data given that the null hypothesis is correct (= P(observed Data | null Hyp)).

We will discuss statistical significance testing extensively in Chapters 9–12, but it is crucial to hammer home (more than once, more than twice) the message that a scientist should not use statistical significance as the primary basis for deciding if a finding provides support for a given hypothesis or is worth reporting.

References

[1] Keller EF. A feeling for the organism: the life and work of Barbara McClintock. W.H. Freeman; 1983.

[2] Hubel DH, Wiesel TN. Receptive fields of single neurones in the cat's striate cortex. J Physiol October 1959;148:574–91.

[3] http://money.cnn.com/2016/06/10/technology/hillary-clinton-google-search-results/.

[4] Lv J, Pan Y, Li X, Cheng D, Liu S, Shi H, Zhang Y. The imaging of insulinomas using a radionuclide-labelled molecule of the GLP-1 analogue liraglutide: a new application of liraglutide. PLoS One May 7, 2014;9(5):e96833. http://dx.doi.org/10.1371/journal.pone.0096833.

[5] Sorge RE, Martin LJ, Isbester KA, Sotocinal SG, Rosen S, Tuttle AH, Wieskopf JS, Acland EL, Dokova A, Kadoura B, Leger P, Mapplebeck JC, McPhail M, Delaney A, Wigerblad G, Schumann AP, Quinn T, Frasnelli J, Svensson CI, Sternberg WF, Mogil JS. Olfactory exposure to males, including men, causes stress and related analgesia in rodents. Nat Methods June 2014;11(6):629–32. http://dx.doi.org/10.1038/nmeth.2935.

[6] Chaudhury D, Walsh JJ, Friedman AK, Juarez B, Ku SM, Koo JW, Ferguson D, Tsai HC, Pomeranz L, Christoffel DJ, Nectow AR, Ekstrand M, Domingos A, Mazei-Robison MS, Mouzon E, Lobo MK, Neve RL, Friedman JM, Russo SJ, Deisseroth K, Nestler EJ, Han MH. Rapid regulation of depression-related behaviours by control of midbrain dopamine neurons. Nature January 24, 2013;493(7433):532–6. http://dx.doi.org/10.1038/nature11713.

[7] Tye KM, Mirzabekov JJ, Warden MR, Ferenczi EA, Tsai HC, Finkelstein J, Kim SY, Adhikari A, Thompson KR, Andalman AS, Gunaydin LA, Witten IB, Deisseroth K. Dopamine neurons modulate neural encoding and expression of depression-related behaviour. Nature January 24, 2013;493(7433):537–41. http://dx.doi.org/10.1038/nature11740.

[8] Lugli G, Larson J, Martone ME, Jones Y, Smalheiser NR. Dicer and eIF2c are enriched at postsynaptic densities in adult mouse brain and are modified by neuronal activity in a calpain-dependent manner. J Neurochem August 2005;94(4):896–905.

A. DESIGNING YOUR EXPERIMENT

6

Power Estimation

INTRODUCTION

I used to think that carrying out **power estimation** was merely some sort of technical procedure that allows one to plan how many observations (items, subjects, datapoints) need to be included in a study. However, looking at the literature, I am struck by the almost total absence of prospective power estimations [1], with only ~3%–5% of articles, at most, reporting this within published articles, apart from a few fields such as clinical trials and genome-wide association studies. The National Institutes of Health and other funding agencies are starting to require seeing power estimation in grant proposals, so that they feel confident that the proposed studies will not be underpowered. Even so, power estimations are not described in the overwhelming majority of published articles. If there is a smoking gun revealing the prevalence of a crisis in scientific practice, this is it! And strategically, if books such as this one can convince students that carrying out prospective power estimation is actually something essential to do, this may produce a ripple effect to improve practice in many other areas as well.

WHAT IS POWER ESTIMATION?

When planning out a study, you would like to know in advance that—whether the results turn out to be positive or negative—you will be able to publish the study and have confidence in your results. On the one hand, this means having confidence that positive effects (e.g., finding that two groups are significantly different from each other) are real and are not simply produced by sampling error. On the other hand, this also means having confidence in negative effects as well. If you find that two groups are NOT significantly different from each other, are you sure that your experiment was sensitive enough to detect differences in the first place?

Power is the probability that if an effect (of a given size) does exist in your data, then you will find that the effect will achieve statistical significance in your experiment.

Certainly this needs to be estimated in advance, during the planning stage. So why do so few scientists do this? There are probably several reasons. One is that power estimation often says that the number of samples or subjects needed is, uncomfortably, much higher than the usual number historically employed in the field. For example, many animal studies typically employ 8—10 animals per group; many published studies of postmortem tissue typically employ 5—20 samples per group; and fMRI studies may employ ~20 subjects per group— whereas power estimation may suggest that 2—3 times as many are needed [2]. Often the design and size of a study is pragmatically limited by the budget or the availability of samples or subjects. Thus, the information given by power estimation is unwelcome when it runs at odds with "usual practice."

However, another reason is that it is not so easy to estimate the sample size needed for an experiment. The investigator is expected to know (or guess) the results in advance, including the effect size, the variance of both control and experimental groups, the pairing scheme (if any), and the shape of the data distributions. Some of these can be known from previous studies or from carrying out pilot studies—and in fact, one of the main reasons to carry out a pilot study is to provide estimates of these parameters so that power can be ascertained reliably. The burden of specifying the parameters (and perhaps the element of fantasy inherent in doing so) may dissuade even well-meaning investigators from carrying out prospective power estimation.

THE NUTS AND BOLTS

Fig. 6.1 shows a diagram of the simplest and most common type of power estimation, illustrating the G*Power 3 open software tool as described by Faul et al. [3] and available for free download at http://www.gpower.hhu.de/en.html.

In Fig. 6.1, we see the distribution of sample datapoints for two groups (it is easiest to think of them as a control and experimental group, although they could be two different experimental groups). Each group is idealized as following a normal distribution, which is plotted in standard form for the control group (see Chapter 3). In the case shown here, both curves have the same standard deviation (SD), although one can specify SDs for each curve separately. We want to be able to detect true differences in means (=the **effect size**) that may exist between the two groups of a certain stated, prespecified size. We might not actually have a good idea how big the effects will be in this experiment, but we do want to be sure that our experimental design is adequate to detect effects in a range that is likely to be scientifically meaningful. So, for power estimation, we prespecify the magnitude of the underlying or "true" effect size that we want to be able to detect. However, we know that due to random sampling, each time when we carry out an experiment, the mean values of each sample will vary randomly (see Chapter 3) and so the observed effect size will vary as well. That means that the t-test will give a different *P*-value each time and will not achieve statistical significance every time (unless the true effect size is extremely large).

Power is the likelihood that if the true effect size is as specified, then when a single experiment is run, the t-test will achieve significance. The power is interrelated with all of the parameters of the experiment:

1. the means of each curve (this specifies a difference between means of a specific size),
2. the SDs of each curve,
3. the type of t-test to be performed (e.g., one-tailed or two-tailed, paired or unpaired, equal or unequal SDs),

A. DESIGNING YOUR EXPERIMENT

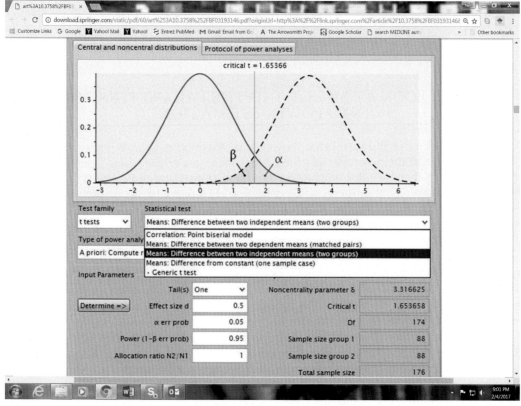

FIGURE 6.1 Screenshot of G*Power 3 software, showing the distribution of two groups that have similar normal shape and variance, but different in mean values. The diagram illustrates how power is calculated (see text). *Taken from Faul F, Erdfelder E, Lang AG, Buchner A. G*Power 3: a flexible statistical power analysis program for the social, behavioral, and biomedical sciences. Behav Res Methods May 1, 2007; 39(2):175–191 with permission.*

4. the desired threshold for deeming a finding statistically significant (in most cases, this is fixed at 0.05, which corresponds to a 5% chance that the finding will be a false positive),
5. the desired power (in most cases, this is fixed at 0.8),
6. the number of datapoints in each group.

Because the parameters are interrelated, you can calculate the value of any parameter if you specify all the remaining parameters. In particular, if you specify 1 through 5, this determines the minimum number of samples or subjects that you need in each group. And that, in a nutshell, is what the goal of power estimation is—to learn how many subjects or samples you need. Conversely, if you specify a given number of datapoints in each group, you can calculate what the resulting power of those experiments will be. (G*Power software can also be applied when experimental designs are more complex and for a variety of statistical tests.)

A **powerful** experiment is one that is able to detect small effect sizes with high confidence. Effect sizes are usually stated in SD units (often referred to as **Cohen's d**): A small effect is one where the experimental mean lies ~0.2 SD away from the control mean, a moderate effect is

~0.5 SD from the control mean, and a large effect is ~0.8 SD away from the control mean. It is not always important to detect small effects, of course. This depends on the nature of the experiment and the underlying science. But an **underpowered** experiment may not even be able to reliably detect moderate to large effects!

A CLOSER LOOK AT FIG. 6.1 AND THE PARAMETERS THAT GO INTO POWER ESTIMATION

Look at the left curve in Fig. 6.1 that shows the distribution of sample points in the control group. If the mean of the experimental group is greater than a critical value, falling to the right of the vertical line, then the t-test will be said to achieve statistical significance. (We will discuss the factors that determine the exact position of the vertical line in Chapter 9.) But remember that repeated sampling from the control group can also produce mean values of that size or greater—this is shown in the diagram as the portion of the control curve that lies to the right of the vertical line (this is called α). This parameter, α, is the probability that an observed mean difference of that size or greater will be a **false positive**, that is, the experimental group is actually being sampled from the same population as the control group. The value of α is generally set by the experimenter at 0.05, meaning if one observes an effect size just at the vertical line, at the threshold for achieving significance, then there is a 5% chance that this is a false-positive finding. A false-positive finding is also known as a **Type I error**.

Note that even if the experimental group has a true difference in means from the control group, generally there is some overlap between the two curves. Conceptually, each time the experiment is run, the observed experimental mean will be somewhat different, and sometimes it will fall to the left of the vertical line (see Fig. 6.1). The area of the curve to the left of the vertical line, called β, gives the probability that a true effect size will NOT achieve statistical significance, that is, the probability that a true finding will be a **false negative**. A false-negative finding is also known as a **Type II error** and is closely related to power; in fact, power $= 1 - \beta$.

HOW TO INCREASE THE POWER OF AN EXPERIMENT

All things being equal, increasing the number of samples or subjects N in each group will increase the power. [Increasing N decreases the SEM ($=$ sample SD/$\sqrt{N}$) for each curve, which shrinks the 95% confidence interval of the estimate of the mean.] However, anything that you can do to decrease variability within the control and/or experimental groups will have the effect of increasing power (Fig. 6.2). This includes experimental design strategies such as pairing subjects (e.g., pairing each control with a matched experimental subject or using an individual as its own control), using stratification, or choosing more sensitive or precise assay methods. One can also increase power by improving data cleansing strategies, such as more careful inclusion of subjects (via thresholding or removal of outliers), binning of outcome values, or better normalization of outcome values. Finally, it is always a good idea to examine previous studies, and pilot studies when available, to make a more informed "guess" of the parameters used for power estimation. This will help ensure that the power of the experiment more truly corresponds to what you expected at the outset.

FIGURE 6.2   **More power is generally better.**

WHAT IS THE POWER OF PUBLISHED EXPERIMENTS IN THE LITERATURE?

Publications rarely mention carrying out prospective power estimation (and even more rarely do they specify the full set of parameters that they entered into the software!). However, a number of investigators have estimated the power of experiments reported in a number of biomedical fields. At this point in the book, you probably will not be shocked to learn that the results were shocking.

Remember that a power of 80% is the standard benchmark for scientific studies. A study of psychotherapy articles found an average power of 49% [4], whereas neuroscience studies averaged power of 8%–22% [5] and in psychology, around 37%. For a true effect that is relatively small (0.2 SD), the estimated power of experiments in the social and behavioral sciences was about 24% [6], and these studies all agree that power has been getting no better during the past six decades. Together with the evidence that the majority of experiments cannot be successfully replicated (see Chapter 1), this indicates that the failure to carry out prospective power estimation, and to actually employ the sample sizes that are deemed necessary, is a very serious problem.

THE HIDDEN DANGERS OF CARRYING OUT UNDERPOWERED EXPERIMENTS

One might think that the biggest effect of carrying out underpowered experiments is the loss of ability to detect true effects. That is, underpowered experiments fail to detect many findings that are sitting in plain view within the data. Yet, ironically, there is no plethora of papers reporting negative findings, and in fact, the vast majority of published experiments report positive findings. What does it mean when an underpowered study reports something positive? Would you not imagine that these articles would only report a subset of experiments (i.e., those that had positive outcomes with larger effects) and so be especially reliable?

On the contrary, and somewhat paradoxically, an underpowered study is *more* likely to report inflated effect sizes, and *more* likely to report false-positive findings [5,6]. Think of it this way: If you have only 5 samples per group, it is pretty easy to observe random fluctuations that produce a large mean difference between groups simply by chance; in contrast, if you have 50 samples per group, fluctuations in a few samples will have little overall effect. And since only the observed large effects will be deemed significant in an underpowered study, those are the ones that will be reported. This may be the worst single consequence that low power has wrought upon the scientific literature.

THE FILE DRAWER PROBLEM IN SCIENCE AND HOW ADEQUATE POWER HELPS

The file drawer problem in science refers to the fact that scientists are loath to write up and publish negative findings (Fig. 6.3). From the perspective of an individual investigator, this makes a lot of sense:

1. Most negative results are less important, and spending time writing them up is time lost from carrying out new (hopefully more exciting) experiments.
2. There are few positive incentives for publishing null results—they will often not be accepted by leading journals, and at best will be a line on the CV.
3. And there are definite negative incentives—one may get the reputation for publishing unexciting work.
4. If the null findings disagree with the previous published results of others, you run the risk of getting entangled in fights that will make no one happy.

FIGURE 6.3 **The file drawer problem.** Scientists need plenty of drawers to stick those studies that are never written up for publication – particularly those that generate negative findings.

A. DESIGNING YOUR EXPERIMENT

A recent model suggested that scientists optimizing their incentives will carry out experiments with 10%–40% power, resulting in half of the findings being false positives [6,7]. Unfortunately, the lack of negative findings makes the literature unbalanced, and skewed overall toward positive findings. This is one (though not the only) reason that the literature shows an excess of papers that report positive findings beyond what is expected given the inherent power of the experiments [8,9]. As we already noted, positive findings that arise from underpowered studies are likely to be false positive, and indeed, a consortium attempting to reproduce findings reported in the psychology literature found that only about one-third could be replicated [10,11].

How would adequate power help solve the file drawer problem? It will not directly alter the incentives that tempt/prod/direct scientists to favor positive findings, but there is at least one way that it would help: At present, when an investigator observes a null finding, he or she is likely to disbelieve the results. After all, to publish a negative finding, one must be able to demonstrate that the experiment had the power to detect a finding had it been there! (No such criterion needs to be satisfied for a positive finding, whose t-test appears to speak for itself.) If the experiment has adequate power, a negative finding will be as persuasive as a positive finding, and the investigator can write it up with confidence.

WHY NOT CARRY OUT POWER ESTIMATION AFTER THE EXPERIMENT IS COMPLETED?

It is much more common to see post hoc power estimation calculated in published papers. That is, after the experiment is complete and the actual parameters are known, the authors plug these into the power estimation and supposedly validate that the experiment did have adequate power after all. Unfortunately, post hoc power estimation is no more than a form of self-deception. Not only does it come too late to guide experimental design, but also the post hoc estimate of power is fully determined by the observed P-value of the statistical test. That is, P-value and post hoc power are two equivalent ways of stating the same thing. If the test achieved significance in comparing two groups, then post hoc tests will always say that the power was adequate. This post hoc procedure is based on the mistaken assumption that the effect size that was observed in the experiment is indeed the true effect size that exists in nature.

Tip: Stay away from post hoc power estimation altogether!

References

[1] Tressoldi PE, Giofré D. The pervasive avoidance of prospective statistical power: major consequences and practical solutions. Front Psychol May 28, 2015;6:726. http://dx.doi.org/10.3389/fpsyg.2015.00726.

[2] Murphy K, Garavan H. An empirical investigation into the number of subjects required for an event-related fMRI study. Neuroimage June 2004;22(2):879–85.

[3] Faul F, Erdfelder E, Lang AG, Buchner A. G*Power 3: a flexible statistical power analysis program for the social, behavioral, and biomedical sciences. Behav Res Methods May 1, 2007;39(2):175–91.

[4] Flint J, Cuijpers P, Horder J, Koole SL, Munafò MR. Is there an excess of significant findings in published studies of psychotherapy for depression? Psychol Med January 2015;45(2):439–46. http://dx.doi.org/10.1017/S0033291714001421.

[5] Button KS, Ioannidis JPA, Mokrysz C, Nosek BA, Flint J, Robinson ESJ, Munafo MR. Power failure: why small sample size undermines the reliability of neuroscience. Nat Rev Neurosci 2013;14(5):365—76. http://dx.doi.org/10.1038/Nrn3475.

[6] Smaldino PE, McElreath R. The natural selection of bad science. R Soc Open Sci September 21, 2016;3(9):160384.

[7] Higginson AD, Munafò MR. Current incentives for scientists lead to underpowered studies with erroneous conclusions. PLoS Biol November 10, 2016;14(11):e2000995. http://dx.doi.org/10.1371/journal.pbio.2000995.

[8] Ioannidis JP, Munafò MR, Fusar-Poli P, Nosek BA, David SP. Publication and other reporting biases in cognitive sciences: detection, prevalence, and prevention. Trends Cogn Sci May 2014;18(5):235—41. http://dx.doi.org/10.1016/j.tics.2014.02.010.

[9] Francis G, Tanzman J, Matthews WJ. Excess success for psychology articles in the journal science. PLoS One December 4, 2014;9(12):e114255. http://dx.doi.org/10.1371/journal.pone.0114255.

[10] Open Science Collaboration. PSYCHOLOGY. Estimating the reproducibility of psychological science. Science August 28, 2015;(6251):349. http://dx.doi.org/10.1126/science.aac4716. aac4716.

[11] Dreber A, Pfeiffer T, Almenberg J, Isaksson S, Wilson B, Chen Y, Nosek BA, Johannesson M. Using prediction markets to estimate the reproducibility of scientific research. Proc Natl Acad Sci USA December 15, 2015;112(50):15343—7. http://dx.doi.org/10.1073/pnas.1516179112.

A. DESIGNING YOUR EXPERIMENT

This chipmunk does not need to carry out power estimation—it can never have too many nuts!

GETTING A "FEEL" FOR YOUR DATA

7

The Data Cleansing and Analysis Pipeline

STEPS IN DATA CLEANSING

Examining the Raw Data

The experiment has been planned and executed, and you are sitting with a pile of **raw data**. What do you do now? It is very tempting to run a statistical test first, to see if any groups are significantly different from each other. But that should be the last thing that you do, not the first.

Data cleansing, the pipeline of steps that transform the raw data to a processed form that is ready for serious analysis, seems deceptively simple and straightforward, but the results of your study depend critically on how you perform these steps. Data cleansing is no less important than experimental design or statistical analysis, since depending on how you carry it out, the same raw data can point to a small effect, a large effect, or an effect in the opposite direction entirely.

No single recipe or list of tasks will apply to all circumstances, but I suggest that the first thing to do when given a spreadsheet of raw data is to plot the distribution of the raw data, both pooled together and in separate partitions. Plot the controls and experimental groups separately—but juxtaposed so you can compare them, say by plotting each group with different colors. Plot different experimental runs separately (but juxtaposed so you can compare them). Look at different individual subjects, different sites (if a multicenter study), and so on. These plots will give you a feel for the overall shape, scope, and variability of the data, as well as detect possible problems with the experiment.

Specifically, by plotting and examining the data, you should ask a set of basic questions that will guide the cleansing process: Are the data distributions at least quasinormal in shape? Bimodal? Strongly skewed with a long tail? Are the baselines consistent in the controls from day to day? Do some groups show high variability relative to others? As discussed below, are there any weird samples, subjects, or groups? Outliers? Missing data? Floor or ceiling effects?

Tip: Proper data cleansing should serve to reduce the amount of variation within individual groups. As a result, the power of the experiment will improve, that is, the number of datapoints that achieve statistical significance (if any) is likely to increase.

Data Literacy
http://dx.doi.org/10.1016/B978-0-12-811306-6.00007-5

Visualization Tools—How to Choose and How to Use

Many, if not most, types of statistical software have plotting capability. However, personally, I generally create simple plots directly within Excel. (Although Excel is opaque to figure out, there are Youtube tutorial videos to cover every common situation.) As an alternative, I have found Weka data mining software [1] (http://www.cs.waikato.ac.nz/ml/weka/) to be easy to install and use (and is free). One can export data from Excel in comma separated (CSV) format and upload it into Weka Explorer with a few clicks. By clicking the Visualization tab, each column of data is automatically plotted against every other one. Data scientists often visualize data using Matlab (a popular commercial software) or R plotting packages (which are open source and free; https://cran.r-project.org/) [2]. Many other types of visualization software have been created for scientists, but these seem to be designed for advanced purposes or advanced users.

Plot and Examine the Data Distributions

Knowing the shape of the data distributions is important for several reasons:

1. First, this will influence the choice of statistical tests to employ. If the datapoints are roughly centered around a single, well-defined peak, we say that this is **quasinormal** (i.e., if we carried out a test for normality, it would pass) (Fig. 7.1), and it is likely that parametric statistics such as t-test or ANOVA can be utilized.

FIGURE 7.1   "No, I said quasi-NORMAL!".

B. GETTING A "FEEL" FOR YOUR DATA

2. Data distributions with two distinct peaks (**bimodal**) or a stretched-out appearance are likely to represent a mixture of different subpopulations having different responses, and this should be analyzed in detail—possibly the subpopulations should be treated separately and not as one big group.

3. If the datapoints cover several orders of magnitude (i.e., the highest datapoint is hundreds or thousands as times as large as the smallest datapoint), or if the data are strongly skewed (i.e., have a long tail at one end), these datapoints can often be transformed to make the curves quasinormal, which permits the use of parametric statistics without any loss of information. Parametric statistics are desirable to use when possible, in most cases, because they are more powerful than nonparametric tests. Also, when measurements are conducted on a log scale (e.g., when measuring pH units), the arithmetic mean of the data is not the most accurate or robust way to summarize the data distribution—instead, the **geometric mean** or **geomean** should be used (Chapter 3).

One popular way to transform datapoints to make the overall distribution more quasinormal is to convert each datapoint to its log (usually chosen as either natural log, i.e., log to the base e, or the log to the base 2 or 10) (Fig. 7.2). This runs into problems when some values equal 0, since log0 is undefined. Alternatively, if the datapoints are all nonnegative (i.e., 0 or greater), one can transform each datapoint to its square root. Both of these methods have the effect of "compressing" extended distributions into a much smaller range.

Look for Weirdness in the Data

Weirdness is not the same thing as outliers (see below). Suppose one of the subjects in your study gave a blood sample that was measured for a panel of RNAs, and all of the RNA values in that sample are ~30% of the values seen in most other samples. No single RNA value

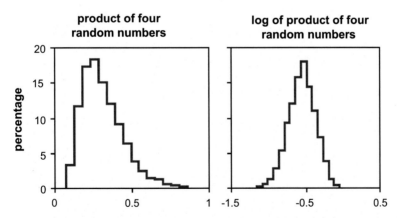

FIGURE 7.2 **Data before and after log_{10} transformation.** Both panels show the frequency distribution for the product of four numbers, with each number having a uniform random distribution between 0.5 and 1. The graph on the left shows the untransformed product; the graph on the right is the distribution of the log-transformed products. Note that the raw data are skewed but the transformed data appear to be roughly normal. *Reprinted from http://udel. edu/~mcdonald/statcentral.html with permission.*

might be an outlier, yet clearly the subject's blood sample is suspect. Samples can also be weird even when their measured values seem OK—for example, if most samples are extracted to yield 1 µg of total RNA and one sample yields only 0.05 µg, it is weird, even if both give similar measured values of specific RNAs. One would not necessarily discard the sample having a low yield—maybe the handling of that sample was done differently, or maybe that sample had high blood lipid content, which affected the efficiency of RNA extraction; you may not know the cause, but it is important to flag that the sample is "weird" before proceeding further (Fig. 7.3).

Similar issues happen with postmortem tissue samples, some of which may show evidence of degradation and should be removed from consideration, even if the individual values measured in the samples are not divergent enough to be called outliers. Another nonbiological example arises when analyzing online movie reviews. Suppose certain individuals tend to give unusually high or low ratings overall, such as all 0s or all 5s. Given the high variability of movie ratings, neither a rating of 0 or 5 would be outliers, yet these individuals may be deemed anomalous and either excluded from the analysis or studied further as separate cases.

Summary Variables

It is good to look at some of the **summary variables** for the raw data, notably the means, medians, and standard deviations, even though these are likely to change substantially after data cleansing has been accomplished. At this stage, one of the main questions is whether the means and SDs are similar across different runs in the **same** group. That is, are the baselines stable within the experiment? Were there overall changes occurring over time (or in particular runs), which might indicate that the assay or the equipment was not well controlled? Baseline values can be thought of as a kind of positive control and should be consistent within different parts of the same experiment.

FIGURE 7.3   **If I removed all the weird datapoints from my study, I would have none left!.**

Another important question is whether the overall extent of variability is similar in the controls versus the experimental groups. Sometimes a change in the amount of variability is itself the primary finding of a study (Chapter 8). More often, if one group shows much more variability than another, it may hint that there is heterogeneity within that group, which should be analyzed further. Heterogeneity may reflect problems with the experiment or may be scientifically interesting. For example, suppose I am interested in relating the incidence of violent crime to mean annual income across different countries around the world. The incident of crimes might be the highest, not in the countries that have the highest mean incomes, but in those which have the highest VARIABILITY in incomes (i.e., the greatest disparity between rich and poor).

High variability can also be caused by the presence of outliers. Note that the standard deviation (SD) is not the best measure of variability to use when comparing different groups against each other, since a distribution that has a larger mean would naturally be expected to have a larger SD. Rather, one should examine the **coefficient of variation** (CoV = SD/mean) to see if any group has a much larger CoV than the others.

Tip: Unless the variability in the control group is low and well controlled, the experiment may not be sensitive enough to detect any differences that may exist in the experimental group. Verify that your negative and positive controls worked properly, before getting too far into the data analysis!

From Floor to Ceiling

Much of science is driven by the desire to measure very low quantities—for example, to sequence individual DNA molecules, to measure intracellular free calcium levels within a single dendritic spine, or to measure the sensitivity of a moth to detect pheromones. Low numbers are important in social science too—the US election of 2000 hinged on accurately counting votes from a few precincts in Florida. Most measurement methods have a limit of sensitivity, below which the signals become buried in noise and uncertainty.

Most experiments in the biomedical and social sciences collect data whose observations are above the detection limit of the measuring apparatus. However, even in these cases, it is generally a good idea to (1) determine exactly what the reliable detection limit is in that particular experiment and (2) set a threshold excluding all smaller datapoints. It is worth seeing if the trends in your smallest datapoints show anything interesting, unique, or divergent from the trends observed overall—if so, you may wish to keep the very low datapoints and analyze them separately. In most cases, however, it is preferred to remove those points that may primarily represent noise instead of meaningful signal. One way to define the reliable detection limit is to plot all datapoints in the experiment binned according to their average values, and the corresponding (CoV). As the average values get smaller and smaller, there will be a point at which the coefficient of variation CoV abruptly starts to rise, which can be chosen as the threshold.

Ceiling effects occur when the measuring devices (or the system under study) become saturated or maxed out. For example, a common practice among people who perform Western blots is to compare samples that are sitting on adjacent lanes on the same blot and measure the relative intensity of immune staining for some protein of interest (Fig. 7.4). The ratio of intensities is not a reliable quantitative estimation of the relative

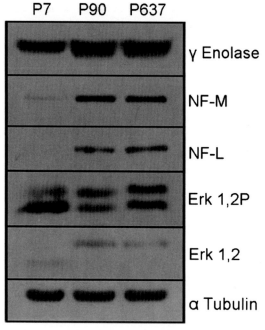

FIGURE 7.4   **The validation of some proteins, differentially expressed with age, was performed by Western blot.** Mitogen-activated protein kinase 1, 2 (Erk1,2), phosphomitogen-activated protein kinase 1, 2 (Erk1,2P), neurofilament low (NF-L), neurofilament medium (NF-M), and gamma enolase (γ Enolase) in the olfactory bulb between the different development stages (P7, P90, P637). α-Tubulin staining was used to ensure the equal loading of proteins. *Reprinted from Wille M, Schümann A, Kreutzer M, Glocker MO, Wree A, Mutzbauer G, Schmitt O. The proteome profiles of the olfactory bulb of juvenile, adult, and aged rats—an ontogenetic study. Proteome Sci February 15, 2015;13:8. http://dx.doi.org/10.1186/s12953-014-0058-x with permission.*

amounts of protein present, however, unless you know that the intensity increases linearly with the amount of protein loaded. Also, when comparing different adjacent lanes for band intensity, the different lanes need to be loaded with the same amount of protein, which is usually tested by reblotting the same blot to detect some abundant constitutive protein such as beta-actin or alpha-tubulin.

Another example of a ceiling effect is a clinical study asking if a new type of bandage improves healing of scraped knees when assessed 21 days later. If all of the control knees have healed at 21 days, then it would be impossible to detect any effects due to the bandage. A better design might choose a better endpoint (maybe 7 days?), different subjects (maybe examine people with immune deficiencies who do not heal well on their own), or include different relevant outcomes (such as the prevention of scars).

Outliers

These are datapoints that are very different from the vast majority of the points collected in an experiment (Fig. 7.5). There are no universally agreed criteria for calling a datapoint an

FIGURE 7.5   **An outlier.** *Courtesy of Jessica Tam, https://commons.wikimedia.org/wiki/File:Smile_2.jpg, with permission.*

outlier, but I personally like to say that an outlier is a datapoint that lies three or more SD away from the mean of the group that it is in. (If the data distribution is highly skewed, use the median value instead of the mean.)

Outliers may represent frank errors—for example, the measuring apparatus might have been malfunctioning due to a power surge—or even due to transcription errors, as when a person forgets to put a decimal point when writing the number in a notebook. They may represent samples that were weird for some reason; for example, when studying survival of kidney transplants, some kidneys may have been used that were damaged or handled or subjected to delays, and these may produce unusually bad or variable outcomes. Such outliers should be assessed and removed, if possible.

Outliers can also represent heterogeneity within the group being studied, and in such cases, it is less clear whether they should be removed or studied separately. For example, if I am studying the factors predictive of risk of drivers dying in car crashes, and I sample randomly from everyone in the entire population, the strong factors are likely to relate to age and experience of the driver, and amount of alcohol consumed. But the sample may include a few professional NASCAR drivers, long-haul truck drivers, and elderly codgers with bad vision, each of whom are strongly divergent from most other drivers.

When the raw data are plotted and examined, one should flag the outliers and assess them carefully. For example, are outlier values restricted to just one experimental group or just to the control group? Are all of the outliers associated with one particular sample? Are the outliers mainly found below the mean? Are they all very high values? Or scattered randomly about? Are there only one or two outliers in a group, or do they represent a significant number of points (even if a small minority of the total)?

Sometimes the outliers are themselves the most interesting story in an experiment. Suppose I am studying the relationship between the amount of money earned by lottery winners and their later quality of life. Most of the datapoints will be $0, some $5, or $30. Only one or two will earn very high winnings, e.g., $30 million dollars. These few are outliers based on statistical criteria, yet if I removed them, I would have a totally incomplete view of the situation. To give another example, if I am studying 100 rats, and they all weigh roughly 200 g, I am likely to remove data for one rat that weighs 600 g! However, if there are three rats (especially all in one treatment group) that weigh 600 g, I would not remove them automatically, but rather would examine what this means. Severe weight gain caused by the experimental intervention would be important to document and analyze, even if it only affects a small minority of the rats in the overall study.

It is essential, when reporting an experiment, to state how you define outliers, to describe the outliers that were found, and to analyze the data both with and without including the outliers.

Tip: If your overall conclusions are different when including versus excluding outliers, that raises a big red flag, and you do not have a very robust finding in the first place!

Missing and Censored Data

Missing data can arise for many reasons. Measuring equipment might have failed, or the measuring assays might be flawed. (I recently did a microarray experiment using a commercial 384-well microassay plate, where one specific well gave a false reading of 0 for all samples.) The experimenter might even have run out of money before certain measures can be done for all samples or subjects! There can be errors in data entry. There can be incompatibilities when translating from one system to another, such as when combining data across two hospitals that have different ways of encoding diagnoses. One hospital may have a code for "developmental delay, not otherwise specified" that the other hospital does not, resulting in measures that cannot be applied to all samples (unless it is possible to map that diagnosis unambiguously to a corresponding diagnosis used in the second hospital).

One popular approach to handling missing data is simply to remove samples or subjects from the study if some of their values are missing. This is the easiest approach, but generally not the best [4]. Another popular approach is to **impute** missing data—that is, to enter the most likely value of the datapoint (usually, the mean value of the group that the datapoint belongs to). Other, more elaborate techniques are also used, but keep in mind that there is no one right answer that applies to all situations. The experimenter needs to exercise judgment, consider the possible causes of the missing data, and weigh the possible impact of handling the data different ways on the conclusions of the study. Above all, when writing up the results, discuss the nature of the data issues, how you dealt with them, and what impact they have on your conclusions.

In contrast, **censored data** refers to datapoints whose values are not entirely known. For example, if I am weighing subjects, and my scale only goes to 300 pounds, then I would not have accurate weights for people who are heavier than that. In a clinical trial, some subjects may drop out of a study early (or during follow-up), resulting in data that may be missing, censored, or both. It is imperative to consider dropout rates in a study because this may be an important clue (maybe even the most important finding of the entire study!). For example, if I am testing a drug versus placebo, and the majority of subjects getting the drug drop out, that is an important clue that there may be adverse side effects.

DATA NORMALIZATION

Normalization is like driving a car—everyone does it, and everyone thinks they are an expert! Yet ask someone if they can drive a stick shift—or change a flat tire—or drive a truck—or drive their car through fog or mud… and their expertise turns out not to be actual proficiency (which requires flexibility and judgment) but merely the ability to go through a habitual set of actions.

One of the purposes of normalizing raw data is to put the data on a meaningful footing. For example, if I inject rats with 10 μg of some drug, the actual dose will be twice as high in a 100 g rat as it is in a rat weighing 200 g, assuming it gets distributed evenly throughout the body. Therefore, it makes sense to analyze the data in terms not of the amount of drug, but the normalized dose in micrograms injected **per gram weight** of the rat.

Another purpose for normalization is to minimize spurious or irrelevant sources of variability in the experiment. For example, if I do a study on cells growing in culture and harvest the cells to measure molecule X, there will be some variation from dish to dish in the total number of cells present per dish, as well as some variation in exactly how efficiently I harvest the cells. Instead of scoring the amount of molecule X obtained per dish, it makes sense to normalize the data to give the amount of molecule X per cell.

Like everything else related to data cleansing, there is no single right way to carry out normalization. Yet your choice of normalization may well determine whether you detect any interesting effect at all, or whether your experiment has a positive or negative outcome. In the example of measuring molecule X in cultured cells, the best way to normalize depends on the biology of molecule X. Do I really want to normalize the value of molecule X to the number of cells in the sample? In some cases, I might want to normalize instead to the average cell volume instead. Suppose that molecule X is a constitutive structural protein in the cytoplasm. If my experimental condition makes cells might grow to twice the volume of the control cells, then each cell will have twice the amount of molecule X just because the cells are larger (and have twice as much of every cytoplasmic molecule), not because there was any specific upregulation or change in molecule X per se. In such a case, I probably want to normalize the amount of molecule X to the amount of some other structural cytoplasmic protein such as actin measured in the same sample. In that manner, the measure will not be sensitive to changes in cell size and will instead tell me how molecule X is being regulated DIFFERENTLY from other cytoplasmic structural proteins.

Several standard strategies for normalizing raw data include the following:

1. Quantile (or rank) normalization

This means converting the raw data to ranks. For example, in an experiment scoring the number of animals swimming across a stream on each of 10 days, we could present the number of animals on each day (raw data) or we could rank each day in terms of highest to lowest number of animals and score the rank. That is, the day having the fewest animals is given rank 1, and so on. Ranks are sometimes presented in terms of percentiles, e.g., a person is in the top 1% of all students on an exam. Ranks are very useful for analyzing data distributions that are not normally distributed, and are used to prepare data for nonparametric statistics (Chapter 12).

2. Mean or median normalization

This means normalizing the raw data to the overall mean or median of some population. Depending on the context, this might mean dividing the raw value of each datapoint by the mean or median of its own group or by the pooled grand mean computed across all groups. Normalizing to the mean is preferred if the population follows a normal distribution, whereas the median is preferred if it does not, or if you are unsure.

Mean normalization is very popular, especially when experimenters do not have absolute baseline values to compare groups against, so they look for relative changes instead. Note that one loses the ability to detect absolute changes of any sample, but instead will detect changes in a sample RELATIVE to the population as a whole. For example, suppose I am measuring microRNA expression levels in some cultured cell type exposed to a drug. Suppose ALL microRNAs go up equally in this experiment. The raw data will show an increase in all values, but so will the mean value, so the mean or median normalized data will show no change at all. If certain microRNAs go up MORE than the average amount, the mean-normalized data will go up, but if certain microRNAs go up LESS than the average amount, the mean-normalized data will show an apparent DECREASE. This is a great example of how normalization schemes can affect not only the amount but also the apparent direction of effect that is scored in an experiment.

A different type of mean normalization is appropriate when control and experimental groups are being compared in an assay whose baseline changes from day to day. In such a case, it makes sense to normalize the datapoints in the experimental group by the mean value of the control group measured on the same day.

3. Endogenous or exogenous standards

Endogenous standards are molecules or entities that are naturally included in the system being studied, and which do not change during the experiment. For example, as mentioned above, one might normalize data to the number of cells, or to the amount of actin, in the sample.

An **exogenous standard** is something added to a sample (or to an experiment) to provide an absolute calibration point. This can be a way of double checking the reliability of the machines doing the measurements; it can also take into account any changes in measured values that are due to the way the samples are prepared.

Endogenous and exogenous standards are not restricted to biochemistry, of course, but can be utilized in many types of experiments. Suppose I am profiling the books written by different authors to best capture their unique styles, and I hypothesize that they may differ in how often they use pronouns. To compare the pronoun frequencies across different authors, it might make sense to normalize the pronoun frequencies by some other endogenous feature, such as how often they use nouns, or by the average sentence length. Exogenous standards cannot be used retrospectively, of course, but let us imagine that you are trying to identify Twitter tweets that are due to "trolls," based both on the content of the trolls' tweets and the responses that they garner [5]. One could send a series of fake troll tweets as exogenous standards and observe the resulting responses. (OK, the ethics of that are questionable, but you get the idea!)

4. Interpolation or linear regression

This method of normalizing datapoints corrects an observed value to remove the effects of some confounding factor. For example, suppose I am studying factors that determine annual income. But one variable that affects income is age across the life span—very young people are unlikely to make much money, and the same is true for old retired folks. There are many ways to handle this situation in designing a study that studies annual income, but one possible way is to survey the overall mean income at each age and plot this curve over the life span. Then, for any individual, take their age and find the average income I that is expected for their age. Finally, take their actual income I_i and normalize it: $I_{normalized} = I_i - I$. Thus, instead of using their income per se as a measure, you measure how much their income deviates from the average expected for their age.

Tip: Generally, it is good practice to analyze your data first without carrying out normalization and then to compare several different normalization schemes. Often experimenters will utilize and compare the results of normalizing data by a number of endogenous and exogenous standards, to see which give the most reliable and scientifically relevant insights into the behavior of the system being studied. These criteria should be checked and reported when you write up your paper.

Tip: A suitable normalizer is one that:

1. **shows no change in mean values across groups and**
2. **has low variability within each group.**
3. **When more than one endogenous or exogenous standard appear to be suitable, generally you should choose the one that produces the lowest variability among outcomes or datapoints observed withina the control group after normalization.**

To Bin or Not to Bin?

Binning means to convert continuous raw data into intervals. If your measure is the number of animals that swam across a stream on different days, the raw data are the number of animals on each day, but for some purposes, it may be more relevant or convenient to analyze the data binned in intervals. For example, out of 10 days examined, the number crossing the stream was between 0 and 5 for 3 days; between 5 and 10 for 5 days; between 10 and 15 for 2 days; and between 15 and 20 for 1 day. Binning simplifies the data, highlights natural heterogeneity in your population (e.g., dividing people into "short," "average," or "tall"), minimizes the effects of extreme datapoints, and often makes it easier to visualize large-scale trends when plotting figures.

However, there is a loss of information using binned data, and—important—you cannot know in advance what the optimal way to bin data is in your particular experiment. Is it adequate to divide your population into short versus average versus tall, or is there more insight provided if you divide the group into VERY short, short, average, tall, and VERY tall? If you carry out statistical tests across groups with binned data, you may find that two groups are significantly different using one scheme of binning, but not using another. Putting data into bins of manually determined variable lengths is particularly prone to giving false results and should be avoided (Fig. 7.6) [6]. Thus, binning should be carried out judiciously. Again, make sure that your conclusions are not materially affected by the binning scheme.

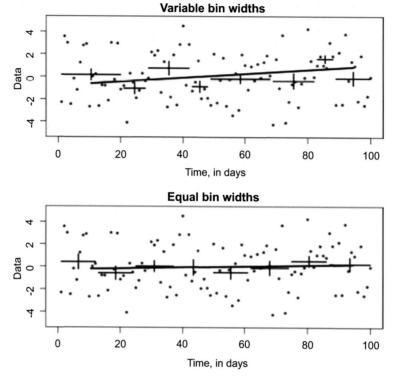

FIGURE 7.6 An example of the significant biases that can result when fitting with variable width bins. Data are generated over 100 days, sampling from a normal distribution with no trend (*red points*). The top plot shows the fit when variable width bins are employed to optimize the agreement with the grouped data (*black points*) to a model that assumes that the slope is 0.02 day−1 day. The fit returns an estimated slope of 0.018 ± 0.006, which is three SDs from the true value. The bottom plot shows the fit to the data grouped into bins of equal width. The fit yields an estimated slope of 0.005 ± 0.007, consistent with the true value. *Reprinted from Towers S. Potential fitting biases resulting from grouping data into variable width bins. Phys Lett B. July 30, 2014;735:146−148 with permission.*

A BRIEF DATA CLEANSING CHECKLIST

1. Plot and visualize the distribution of the raw data, pooled and in separate groups.
2. Make sure that baselines are consistent across assays and runs. Characterize different sources and types of variability that might be reflected in your data. Verify your negative and positive controls.
3. Make transformations (e.g., log or square root) if indicated.
4. Look at summary descriptive statistics in the raw data (means, medians, SDs, and CoVs for each group).
5. Identify and handle ceiling effects, weird data, outliers, missing data, and censored data.
6. Normalize and bin the data as appropriate.
7. Look at summary descriptive statistics in the cleansed data (they will be "for real" now).
8. Think about how to perform statistical testing in the most appropriate manner.

References

[1] Frank E, Hall MA, Witten IH. The WEKA Workbench. Online appendix for data mining: practical machine learning tools and techniques. 4th ed. Burlington, Massachusetts: Morgan Kaufmann; 2016.

[2] Adler J. R in a nutshell: a desktop quick reference. 2nd ed. O'Reilley; 2012.

[3] Wille M, Schümann A, Kreutzer M, Glocker MO, Wree A, Mutzbauer G, Schmitt O. The proteome profiles of the olfactory bulb of juvenile, adult and aged rats — an ontogenetic study. Proteome Sci February 15, 2015;13:8. http://dx.doi.org/10.1186/s12953-014-0058-x.

[4] Rubin LH, Witkiewitz K, Andre JS, Reilly S. Methods for handling missing data in the behavioral neurosciences: don't throw the baby rat out with the bath water. J Undergrad Neurosci Educ January 1, 2007;5(2):A71—7.

[5] Mojica L.G. Modeling Trolling in Social Media Conversations. arXiv:1612.05310v2.

[6] Towers S. Potential fitting biases resulting from grouping data into variable width bins. Phys Lett B July 30, 2014;735:146—8.

Topics to Consider When Analyzing Data

WHAT IS AN EXPERIMENTAL OUTCOME?

In most experiments, the investigator is examining whether the set of datapoints in the experimental group(s) shows a change in mean (or median) compared with the control group(s). However, it bears repeating that there are at least three other significant types of endpoints that can occur, which you should check for in your experiment:

1. **A change in variance**

   Sometimes the relevant outcome is a change in the amount of variability occurring in one group. For example, suppose I am comparing the response patients to a drug versus a placebo, and suppose the mean response is exactly the same in both groups, but the response variability is much higher in the drug-treated group. This is a sign that some subgroup of patients did respond to the drug particularly well, and it would be interesting to see if those have something critical in common, e.g., share a specific genetic makeup. Another example is the expression of genes in different diseases. Some genes show differential expression (up or down), but others show differential variability instead [1].

2. **Shape of the data distribution**

   In one study, my colleague Vetle Torvik and I were studying a corpus of biomedical terms, which we scored and plotted (Fig. 8.1; see details in Ref. [2]). The data distribution resembled a normal distribution but was stretched out in one direction (Fig. 8.1, panel I). This asymmetry indicated that the set of terms really comprised a mixture of two subgroups that fit overlapping normal distributions (Fig. 8.1, panel II and III).

3. **A change in how different datapoints or outcomes are correlated with each other**

   Imagine that we are studying factors that contribute to success in college (i.e., the probability of graduating after 4 years). We measure the amount of time studying and the amount of time spent in bars for each student. Suppose these two parameters are completely uncorrelated in freshmen, but have a strong inverse correlation in seniors (i.e., the more time they spend in bars, the less time they spend studying). Even if the mean time studying does not change between freshmen and seniors, the change in correlation indicates some change in their habits and coping strategies.

Data Literacy
http://dx.doi.org/10.1016/B978-0-12-811306-6.00008-7

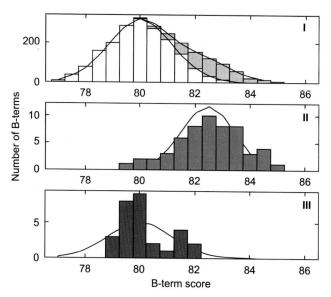

FIGURE 8.1   **An analysis of word feature scores (see Ref. [2] for details).** The data set exhibited a distribution stretched out in one direction (Panel I), which fit a mixture of two overlapping normal distributions (Panels II and III). *Reprinted from Torvik VI, Smalheiser NR. A quantitative model for linking two disparate sets of articles in MEDLINE. Bioinformatics July 1, 2007;23(13):1658–1665 with permission.*

WHY YOU NEED TO PRESENT AND EXAMINE ALL THE RESULTS

Suppose you are looking for an investment counselor to handle your hard-earned savings. One counselor advertises that his fund has increased 200% in the past year, whereas another says that her fund has increased 20%. Who should you choose to manage your money?

This boils down to asking: What data do you need to make an informed decision? At the least, you should learn the size of each investment fund. How long has it been in business, and has it been growing steadily over time? How variable have their returns been, and how consistent relative to the market as a whole? How do they do in good years, and how do they do in a down market? You cannot simply compare the two counselors by looking at differences, ratios, fold-changes, z-scores, or the like. And the same is true when writing up (or reading) a scientific article. You need to analyze and present ALL the results (including the raw and cleansed data sets) to interpret the results in their proper context.

DATA FISHING, P-HACKING, HARKing, AND POST HOC ANALYSES

In Chapter 2 we said that conducting a scientific investigation should be like conducting a court trial—reaching a conclusion only when it seems solid beyond a reasonable doubt—only when all of the other alternatives have been pondered and found to be much less plausible than the preferred interpretation of the findings.

This way of thinking is at odds (no pun intended) with the type of statistical criteria that scientists are generally assumed to follow: That is, if the data show some differences between groups that achieves statistical significance, then scientists tend to conclude that it is a "real" finding and worth reporting—regardless of whether the finding is related to their original hypothesis or research question, or was observed incidentally, and regardless of whether there is any supporting evidence that the finding is real other than the statistical test itself. So what is a red-blooded, keen-observing scientist to do? Ignore patterns in the data just because he or she did not know to expect them in advance? Or what?

Finding a significant difference incidentally should be the impetus to *start* a further investigation. The statistical test is the beginning of a process, not the end. As we will discuss in Chapter 9, null hypothesis statistical tests only tell you that the observed difference is unlikely to be due simply to sampling error. A statistical test does not assure you that the difference is scientifically meaningful; that the sampling was done properly in the first place; or that the effect is free from bias, confounds, or errors.

Suppose I am walking by an alleyway and happen to see three pennies on the ground. They are "real," but what should I make of them? In contrast, suppose I am looking for a missing teen and suspect that he/she has joined up with a gang. Now I see three pennies on the ground—and take them as evidence that a group has recently been pitching pennies there. The same data can be meaningless or highly meaningful, depending on my own purposes.

When scientists are exploring what patterns they can see in their data, it is very common to divide the groups into subgroups in an open-ended manner and see if any of the subgroups show significant differences. For example, if my original research question is whether Northerners eat more fish than Southerners, I might check each Northern state individually against each Southern state—or check Northern males versus Southern males and Northern females versus Southern females—or check each age group separately (teens, adults, elderly)—or check the results for freshwater fish versus ocean fish, and so on. Instead of a simple pairwise test (Northerners vs. Southerners), I can potentially carry out a plethora of pairwise subgroup tests in an exploratory fashion. Assuming that each test has a 5% chance of being a false-positive finding, I am almost guaranteed to find something statistically significant—and false!

This practice, variously called **data dredging**, **data fishing**, or **p-hacking**, is a major source of false-positive findings reported in the literature [3]. It is OK to carry out multiple exploratory tests, provided that (1) you correct the significance level threshold for the total number of pairwise tests performed (see Chapter 11) and (2) you investigate any incidental finding to obtain additional evidence, arising from a new and deeper investigation (not simply using the same data set).

You Carry Out an Experiment, and Get a Single Effect With $P = .07$. What Are You Entitled to Do Now? What Must You NOT Do?

In this case, there is only one statistical test comparing an experimental group versus the control group and the observed difference is just shy of significance. This is another common scenario in which investigators often unwittingly engage in data fishing.

What you *are* entitled to do is to reexamine your raw data and see if better data cleansing and data analysis can be employed to reduce variability within the control group. For example, if your data are pooled across several different experimental runs, possibly it would help block the data so that you only compare each experimental group versus the control group for each run separately. Or, better normalization might be in order.

What you are *not* entitled to do is to run more animals and repeat the statistical test after each animal, or after each set of 10 animals, until the *P*-value is less than 0.05. This is another example of performing multiple exploratory tests and the resulting false-positive rate will be much greater than 5% [3].

Tip: The number of subjects, items, animals, or other datapoints to be tested must be prespecified at the outset of the experiment, as calculated by prospective power estimation.

Yet another type of data fishing occurs when the investigator is not quite sure which outcome measures are most appropriate, so he or she tries out the results of choosing various combinations. For example, suppose the question is whether the US economy does better under Democrats or Republicans. But what do we mean by "economy does better"? Do we count the employment rate, the GDP, the stock market index, the rate of inflation, or any combination of these? This gives potentially 15 different outcomes that could be examined, some of which are likely to favor Democrats and others which are likely to favor Republicans. The FiveThirtyEight website hosts a charming interactive visualization for this example, to see the impact of including different predictors, outcomes, and ways of cleansing the data: http:// fivethirtyeight.com/features/science-isnt-broken/.

Is this scary? A related concept is **HARKing** (Hypothesizing After the Results are Known), in which an investigator writes up his/her paper as if the observed results had been his/her hypothesis all along (Fig. 8.2) [3,4].

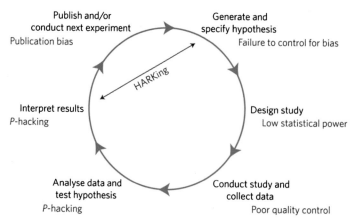

FIGURE 8.2   An idealized version of the hypothetico-deductive model of the scientific method is shown. Various potential threats to this model exist (indicated in red), including lack of replication, hypothesizing after the results are known (HARKing), poor study design, low statistical power, analytical flexibility, p-hacking, publication bias, and lack of data sharing. Together these will serve to undermine the robustness of published research and may also impact on the ability of science to self-correct. *Reprinted from Munafò MR, Nosek BA, Bishop DV, Button KS, Chambers CD, du Sert NP, Simonsohn U, Wagenmakers EJ, Ware JJ, Ioannidis JP. A manifesto for reproducible science. Nat Hum Behav January 10, 2017;1:0021 with permission.*

PROBLEMS ASSOCIATED WITH HETEROGENEITY

When the Population Is Heterogeneous

If I am fishing in a stream, what is the best bait to use? What is the best time of the day to fish? And should I cast the line in mid-current, or in the weeds near the river bank? If I am only looking for a single type of fish (say, a largemouth bass), I can learn from experience and study to optimize my strategy pretty well. But suppose I am looking for whatever fish I can find. When the stream is home to a mix of perch, pike, catfish, salmon, eels, and so forth, no single strategy is likely to work well to catch every type of fish. I may not even know in advance what type of fish I am likely to encounter. And, the fish population itself is not constant, but changes, e.g., depending on the season.

This, in a nutshell, is the problem of population heterogeneity. If I want to ask a scientific question about fish in that stream, I either have to restrict my attention to a single type of fish (in which case I risk getting a skewed and partial perspective) or look for trends that hold over a variety of fish (in which case I risk lumping trends together in a way that is not biologically meaningful). There is a third alternative, of course, which we discussed in the previous section: I can collect the data descriptively for each type of fish separately and then explore how to lump together the different types of fish into categories in a manner that will reveal the most meaningful trends. The problem with the third approach is that I will be proliferating the number of comparisons made and so increasing the risk that any apparent relationships I find will be false positives.

To give another example, if I seek to examine how tax incentives will affect new hiring, there is not likely to be one simple answer. Instead, the answer will vary across the groups that comprise the overall population: blue-collar versus white-collar jobs, urban versus rural areas, and manufacturing versus service sectors, to name a few. A good scientific study will consider, ahead of time, which groups are most important or relevant to study and will take the heterogeneity into consideration.

When the Outcomes Are Heterogeneous

Outcomes will vary, of course. A baseball player will not strike out every time he or she comes up to bat (we hope). If you plot the outcome percentage of hits during a game from a single player over time, you might expect to see some sort of well-behaved curve with a single **mode** or most frequent value. To say that a player's outcome is **heterogeneous** (rather than simply variable) is to say that they follow substantially different underlying curves at different times. For example, he or she might take a performance-enhancing drug on certain occasions and not others. It is not simply that the mean value of their batting average would improve on drug, but the variance and indeed the entire shape of the outcome distribution may be different than when tested on versus off drug. Another example is when a person is vaccinated—they might respond to vaccination (by raising antibodies) one way under normal situations, but if they are vaccinated when they have a cold or fever, they might respond with a quite different time course and different peak response.

Another way that outcomes can be heterogeneous across a population is when *different* subjects follow different outcome curves. The distribution of outcomes, when summed up

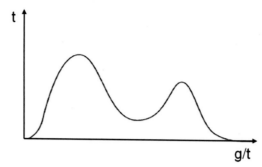

FIGURE 8.3 A typical bimodal distribution. *Reprinted from https://commons.wikimedia.org/wiki/File:Bimodal_geological.PNG with permission.*

across the entire population, might produce a curve that is multimodal (having more than one "hump," see Fig. 8.3) or might be stretched out (see Fig. 8.1).

When different subjects follow different curves, this is most likely a clue that there exists population heterogeneity of some sort as well. For example, if you give a drug to a person, the rate at which the drug is cleared may vary according to their gene profile (particularly the activity of their cytochrome P450 enzymes). However, the rate at which the drug is metabolized may also vary depending on incidental factors such as the time and size of the person's most recent meal.

When the Composition of the Population Changes During the Course of the Study

I want to see if feeding fish in a particular stream early in the season will increase their size at the time of capture. I throw in Capt. Jack's Special Fish Food every other day from March to August and measure the size of fish caught during each month from March to September. The population of fish is 50% bass and 50% minnows during March, and as the bass get caught by fishermen over time, by September we have 10% bass and 90% minnows. **The bass have gotten larger as the year has worn on and so have the minnows. Yet the mean size of the overall fish population has decreased substantially!** This is because minnows are much smaller than bass, and they now comprise a disproportionate number of them.

You might think that a change in the type of fish would be pretty obvious and would not fool an investigator who is measuring the size of the fish population. But consider a slightly different scenario: Suppose fishermen do catch and release bass, keep the ones over a certain threshold, and throw the smallest ones back. This goes on all season. Even though each fish grows, by September, the mean size of the BASS population may stay almost the same. In both cases, there is a paradox: **Each individual fish gets bigger over time, and yet, the overall population can stay the same or even decrease** due to a change in the composition of the population itself.

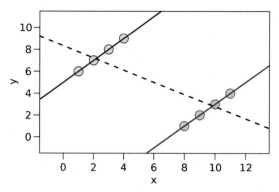

FIGURE 8.4   A graphical example of Simpson's Paradox. A positive trend appears for two separate groups (blue and red), whereas a negative trend (black, *dashed*) appears when the groups are combined. *Reprinted from https:// commons.wikimedia.org/wiki/File:Simpson%27s_paradox_continuous.svg with permission.*

These are examples of **Simpson's Paradox**. As shown in Fig. 8.4, any time a population consists of a mixture of groups A (blue) and B (red), there is a chance that the overall trend across the population will be different than, or even opposite to, the trend within each group.

Simpson's Paradox is a headache for scientists because changes in the underlying population may be overlooked or hard to detect. For example, studies have shown that smokers have a decreased risk of dying from heart disease compared with nonsmokers [5]. Why? Not because smoking protects you against heart disease. On the contrary! However, smokers are more likely than nonsmokers to die of cancer at an early age, i.e., before the peak age for cardiac deaths, so fewer smokers live long enough to be affected by heart disease. The paradox arises because the composition of the population in terms of smokers versus nonsmokers is different at different ages, and the risk of dying from cardiac disease also varies with age (but in a different manner).

The classic example of Simpson's Paradox concerns a lawsuit, which accused University of California at Berkeley of engaging in sex discrimination based on the observation that the acceptance rate for females was significantly less than for males [6]. However, when the data were analyzed separately for each department, it was found that females had **greater** acceptance rates than males across almost all departments. What accounted for the paradox? Females mostly applied to departments that had low acceptance rates, whereas males mostly applied to departments with higher acceptance rates.

Tip: To avoid being fooled by Simpson's Paradox:

First, when you plot your data, examine whether your population consists of subgroups that behave differently from each other.

Second, be aware that overall trends observed across a population may not apply to any or all subgroups within that population.

And third, do not rely simply on statistical thinking to analyze your findings, but use a scientist's eye to consider what mechanisms, driving forces, and causes account for whatever trends you observe.

Looking at an overall relationship is misleading unless one "opens the black box" to see what **mechanisms** account for the relationships between smoking and heart disease, or gender and acceptance rates. Looking at changes in the size of the bass population will be impossible to interpret if one is unaware of the catch-and-release practice.

PROBLEMS ASSOCIATED WITH NONINDEPENDENCE

This is another pervasive (and underappreciated) issue. Whenever you are sampling samples, subjects, or datapoints in an experiment, even randomly, they may not be truly independent of each other. For example, if you are sampling people, even randomly, you might choose several people who come from the same family or live on the same block. (Or, mice that come from the same litter.) Depending on the aims of the study, these interactions might need to be "broken," e.g., by ensuring that no two subjects are first-degree relatives of each other.

The bigger problem comes when the probability of sampling a given subject is related to interactions with other subjects. If I am at a mall asking shoppers to fill out a questionnaire, I am more likely to oversample groups of people who are friends (if not relatives) and likely to share one or more features that might affect the outcomes under study. And of course, simply the fact that I am choosing the subjects at a mall means that I am oversampling people who go shopping (as opposed to shut-ins, online shoppers, or the institutionalized).

Observational experiments are especially prone to the problem of nonindependence. If I am studying whether pork intake is related to risk of heart attack, one approach is to follow a cohort of people who do not eat pork versus eat a lot of pork. But people who do not eat pork differ from those who eat a lot of pork in many different ways, some of which may independently affect the risk of heart attack. For example, people who do not eat pork may have different ethnicities, religions, and lifestyles than those who do. If I do see that pork eaters have different risks, I would not know if it is due to pork per se—and if I do NOT see a difference in risk, I would not know if some of the other correlated factors acted to minimize an inherent risk of pork itself. Any observational experiment needs to identify and control for these correlated factors, or confounds, as much as possible.

Not only samples, but outcomes can also be nonindependent. If the risk of a parent having one child dying in infancy is (say) 1 in 100,000, the risk of having a second child dying in infancy is *not* 1 in 10 billion ($= 1/100,000 \times 1/100,000$). These two events are not independent. The true risk might be as high as 1 in 2, if (say) the parent is a carrier for a genetic disease that is fatal to the offspring.

EVEN PROFESSIONALS MAKE THIS MISTAKE HALF THE TIME!

We will talk about statistical testing in detail in the following chapters, but this point is worth making here: Suppose group A is significantly different from the control group, and group B is *not* significantly different from the control group. Does this imply that A is significantly different from B?

The correct answer is no. Suppose groups A and B have the exact same mean value, but A has a small SD and B has large SD. Depending on the effect size, group A might achieve statistical significance when compared with the control group, yet group B might not. If you want to compare A to B, you need to compare the two of them directly. Seems straightforward, perhaps, but half of reports published in leading journals compared groups A and B the wrong way, by comparing each to the control group and inferring an invalid conclusion [7].

IN SUMMARY

Once your data have been cleansed, you need to have a "feel" for the population being studied, its subgroups, their heterogeneity, and the (possibly different) forces that affect outcomes in each subgroup. Analyzing data is not simply a matter of choosing and performing the "correct" statistical test!

References

[1] Ho JW, Stefani M, dos Remedios CG, Charleston MA. Differential variability analysis of gene expression and its application to human diseases. Bioinformatics July 1, 2008;24(13):i390–3908. http://dx.doi.org/10.1093/bioinformatics/btn142.

[2] Torvik VI, Smalheiser NR. A quantitative model for linking two disparate sets of articles in MEDLINE. Bioinformatics July 1, 2007;23(13):1658–65.

[3] Simmons JP, Nelson LD, Simonsohn U. False-positive psychology: undisclosed flexibility in data collection and analysis allows presenting anything as significant. Psychol Sci November 2011;22(11):1359–66.

[4] Munafò MR, Nosek BA, Bishop DV, Button KS, Chambers CD, du Sert NP, Simonsohn U, Wagenmakers EJ, Ware JJ, Ioannidis JP. A manifesto for reproducible science. Nat Hum Behav January 10, 2017;1, 0021.

[5] Martin A, Martin C. Simpson's paradox: why smoking reduces the risk of dying of cardiovascular disease. Value Health November 2015;18(7):A383. http://dx.doi.org/10.1016/j.jval.2015.09.826.

[6] Kievit RA, Frankenhuis WE, Waldorp LJ, Borsboom D. Simpson's paradox in psychological science: a practical guide. Front Psychol August 12, 2013;4:513. http://dx.doi.org/10.3389/fpsyg.2013.00513.

[7] Nieuwenhuis S, Forstmann BU, Wagenmakers EJ. Erroneous analyses of interactions in neuroscience: a problem of significance. Nat Neurosci August 26, 2011;14(9):1105–7. http://dx.doi.org/10.1038/nn.2886.

It is important not only to analyze the data properly but also to choose the right data to analyze! The investigator has documented an impressive increase in hair gel usage, but missed the fact that the subject has turned into a werewolf.

STATISTICS (WITHOUT MUCH MATH!)

Null Hypothesis Statistical Testing and the t-Test

THE NUTS AND BOLTS OF NULL HYPOTHESIS STATISTICAL TESTING (NHST)

Hypothesis testing is covered in every course in statistics. The t-test is by far the most common statistical test performed by scientists, easily carried out using all statistical software, including Excel. And yet, I find that most graduate students do not have a firm grasp of what the null hypothesis is, nor what hypothesis testing does, nor how to interpret the results of a t-test. I will try to convey the gist of this topic in a practical though deliberately oversimplified way (Box 9.1 fills in more details).

At heart, null hypothesis statistical testing (NHST) is about the baseline or negative control group used in an experiment. It is envisioned as follows:

1. The control group is chosen from a large (generally very large) population of items (say, all the fish in the sea).
2. The experimenter takes a sample containing N items from that population, chosen randomly, such that different items are independent of each other.
3. Each item has a measurement associated with it that is a continuous variable (either interval or ratio measures; see Chapter 4). These measures, or **outcomes**, could be income, weight, longevity, reaction times, and so on, but not simply a label such as "black" or "white" or a movie review with 1−5 stars.
4. The experimenter is primarily interested in estimating the mean value m across the N measurements. A single sample will provide an estimate of the population mean and standard deviation, but may not be very accurate. Instead, it is better to take repeated samples.
5. If the experimenter takes a second random sample from the population, the mean value m_2 is likely to be somewhat different than in the first sample m_1. This is because the items vary among themselves, so that each sample provides a slightly different glimpse or view of the overall population.

127

6. Suppose we sample 3, 4, 5,...k times. The sample means m_1, m_2, m_3,...m_k will themselves form a distribution of values, which has a grand mean $M = (m_1 + m_2 + \ldots m_k)/k$. This grand mean will be a better and better estimate of the true population mean, as the number of samples gets larger.

7. The distribution of sample means, also called the **sampling distribution** or the **"skinny curve,"** will approximate the normal distribution (see Chapter 3). Its standard deviation, also called the **standard error of the mean** (SEM), indicates how much the sample means vary from each other. This reflects both the inherent variability of the items in the population, as well as the uncertainty in estimation due to taking samples.

8. 95% of the sample means will reside within the interval between M minus 2 SEM and M plus 2 SEM. Conversely, 5% of the sample means will fall outside that interval, either lower or higher.

9. As long as the Central Limit Theorem holds (Chapter 3 and see below), it is not necessary actually to carry out repeated sampling to construct the 95% confidence interval for the control group. Given a single sample, which consists of N items and has sample mean m_1 and sample standard deviation SD_s, $SEM = SD_s/\sqrt{n}$ (Chapter 3). The 95% confidence interval will extend approximately between $m_1 - 2\ SD_s/\sqrt{n}$ and $m_1 + 2\ SD_s/\sqrt{n}$.

In a typical experiment, the investigator is comparing a single sample taken from a control group against a single sample of N datapoints taken from an experimental group. (Say, we randomly choose Atlantic bluefin tuna from the Atlantic Ocean as a control group and sample tuna from a particular site of overfishing as the experimental group, asking whether tuna in that site tend to be smaller.) Let us assume that the two groups are really being sampled from the same underlying population. (More precisely, we assume that the control and experimental groups are being sampled from populations that share the same mean, variance, and shape.) This is referred to as the **null hypothesis**. This is very easy to assess.

Fig. 9.1, panel A, shows the skinny curve for the control group. The mean m_2 of the experimental group will fall somewhere along this curve. If m_2 falls within the unshaded region then we cannot reject the null hypothesis. If m_2 falls within one of the shaded regions, then the null hypothesis is rejected, with a stated risk of $P = .05$, $.01$, or $.001$ of making an error (that is, the probability that the experimental group really does have the same mean as the control distribution is .05, .01, or .001). The risk of rejecting the null hypothesis, when the two groups are sampled from populations that have the same means, is called the probability of committing a **type I error**.

Note that we started by assuming that the null hypothesis is true. Therefore, after seeing the results of our experiment, we do not "accept" the null hypothesis, rather, we either reject it as unlikely or we fail to reject it (Fig. 9.2).

BOX 9.1

WHAT MY EXPLANATION LEAVES OUT

1. In practice, everyone uses statistical software to carry out t-tests, and so it is unnecessary to dwell on certain concepts that are usually taught in statistics classes to explain the theoretical underpinning of the t-test. Because I do not assume that the population follows a normal distribution, I do not assume (or care) whether the control samples or experimental samples follow the so-called t distribution. The only thing that is important to know, really, is that if there are N datapoints in a sample, then the sample SD has $N - 1$ degrees of freedom (not N). Even that is not important to know, since the calculator does that for you.

2. A slightly different way to assess the P-value is shown in Fig. 9.1, panel B. Instead of merely asking whether the experimental mean m_2 falls within prespecified shaded areas (as in panel A), a line is drawn at m_2 and the precise P-value is calculated for the dark cross-hatched area to the right of that point. Also, panel B considers the skinny curve constructed for the experimental group. The lightly shaded area of the experimental curve that resides to the left of the line gives the type II error (the probability of having a true difference in means of Δ yet failing to reject the null hypothesis).

3. When you carry out a t-test, the software will display the **t-value**, which is the difference in means Δ divided by the SEM of the control group. A t-value of 2 is approximately at the threshold for statistical significance for a two-tailed test (i.e., corresponds to the limits of a 95% confidence interval), whereas $t = 3$ is approximately at the limit of a 99% confidence interval. The software uses the t-value to calculate the precise **P-value**, which is the area of the control distribution that lies beyond the value of Δ, and which represents the type I error, i.e., the probability of being wrong if one rejects the null hypothesis.

4. Because the null hypothesis assumes that the experimental group will have the same mean, variance, and shape as the control group, the standard t-test only considers the difference in means and ignores the actual variance and shape of the sample taken from the experimental group. However, a more accurate P-value will be obtained by taking into account the variance of both groups (the calculator will take care of this as long as you set the options appropriately; see below).

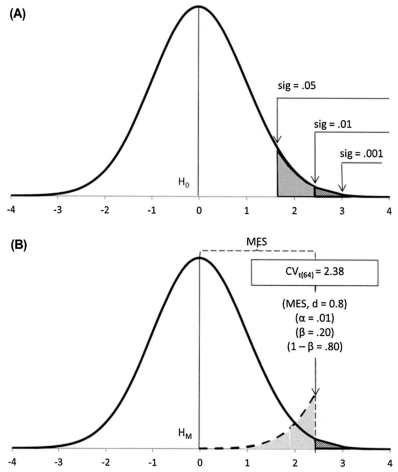

FIGURE 9.1   Two approaches to constructing the sampling distribution, also known as "skinny curve," corresponding to the control group. Shown here is a one-tailed null hypothesis test, in which we test whether the mean of the experimental group is larger than the control mean. In Panel A, if the sample mean of the experimental group falls within one of the shaded areas, the p-value is simply stated as less than 0.05, or less than 0.01, or less than 0.001, as appropriate. In Panel B, given the value of the sample mean of the experimental group, the precise p-value is calculated, and the probability of a type II error (β) is also calculated. See text for details. *Reprinted from Perezgonzalez JD. The meaning of significance in data testing. Front Psychol 2015;6:1293. http://dx.doi.org/10.3389/fpsyg.2015.01293 with permission.*

WHAT NULL HYPOTHESIS STATISTICAL TESTING DOES AND DOES NOT DO

Rejecting the null hypothesis simply says that the observed difference in means between groups is unlikely to be caused by the inherent variability associated with random sampling. This is a very important thing to know, certainly—since if your observed effects COULD plausibly have arisen just as a result of sampling variability, you do not have a robust or solid finding in the first place!

FIGURE 9.2 **Yoda gives some good advice.** *Taken from https://memegenerator.net/instance/68022764.*

But failing to reject the null hypothesis does not provide evidence that the two groups have the same means. Your experiment may not have high enough power, and/or low enough within-group variability, to detect differences that are actually there.

Conversely, rejecting the null hypothesis does *not* establish that a finding is scientifically valid. For example, rejecting the null hypothesis does *not* rule out that your samples might be biased in some manner: a few outliers may have affected the overall trends, your measures might be poor proxies, or your effects might be due to confounding variables other than the ones you were measuring. See Box 9.2 for more myths and debunking.

And there is nothing sacred about using the 95% confidence interval $(+/- 2\,\text{SEM}$ from the sample mean of the control group) as a criterion to decide when a sample mean is "significantly different" from another. As we will see in the next chapter, some scientists feel that it is safer to insist on larger differences before rejecting the null hypothesis such that the type I error is not 5% ($P = .05$), but is as low as 0.1% ($P = .001$).

NHST is perhaps most appropriate for situations such as assessing quality control in industrial production, where the baseline is relatively stable over time and repeated sampling of the same population is actually carried out [2]. Imagine a factory that makes bolts. The diameter of #5 bolts should consistently be of a certain desired size (mean value) and within a certain desired **margin of error** (i.e., so that it falls within the 95% confidence interval of the skinny curve). To check this, one can randomly sample a number of bolts repeatedly and see if their mean value remains within the margin of error. 95% of the times when sampling is carried out, the sample's mean value should remain within the margin of error. If one of the samples does not fall within the 95% confidence interval, then the null hypothesis is rejected (at a 5% chance of being wrong). A typical scientific experiment is different from

---

BOX 9.2

WHAT *P*-VALUES MEAN AND DO NOT MEAN

Suppose you hypothesize that sipping helium causes the pitch of one's voice to be higher. You compare the means of your control and experimental groups (20 subjects in each sample). You perform a paired t-test and get $t = 3$ and $P = .01$, or a type I error of 1%. Which of the following statements follow logically?

1. **You have disproved the null hypothesis.** No, you started with the assumption that the null hypothesis is true—that is what set up the model and math for testing in the first place. So the null hypothesis is never proved or disproved; only rejected or not rejected (Fig. 9.2).

2. **You have proved your hypothesis.** No, you may have strong evidence that there is a difference in means, beyond what can be explained by sampling variability alone. But you have not necessarily proved your hypothesis. For example, suppose the subjects expected helium to raise the pitch of their voices. When they encountered the strange feelings associated with helium, they might have consciously or unconsciously spoken in a higher register. Only after ruling out all plausible alternatives can you accept your hypothesis, regardless of *P*-values.

3. **If you repeated the experiment a large number of times, you would obtain significant results on 99% of occasions.** No, as we saw when discussing power estimation, the probability of achieving significance when there exists a true difference in means is determined by the type II error, not the type I error (Chapter 6). *If* the power is 80% and *if* the true difference in means is actually the same as the difference observed in the experiment, then repeating the same experiment will only achieve significance 80% of the time.

---

this scenario, since it does not have the benefit of seeing the results of repeated sampling, but tries to make inferences from a single data set.

DOES IT MATTER IF MY POPULATION IS NORMALLY DISTRIBUTED OR NOT?

Both the t-test and ANOVA are "parametric" tests, which assume that the population under study follows a normal distribution. When those assumptions hold, parametric tests are more powerful than nonparametric tests (Chapter 12)—or stated another way, it is easier to achieve significance using a parametric test than a nonparametric test. Parametric tests use information about the specific values that datapoints take, in contrast to nonparametric tests that only use information about their relative rank ordering. So there is some incentive to use the t-test or ANOVA whenever possible, even when the assumptions of normality do not strictly hold.

The Central Limit Theorem (Chapter 3) states that if sampling is repeated many times, then the skinny curve will approximate a normal distribution. However, even when sampling is carried out only once taking N datapoints per group, we can construct a skinny curve that follows a normal distribution:

$$\text{Skinny curve mean} = \text{the sample mean}$$
$$\text{Skinny curve SD} = \text{the sample SD}/\sqrt{N} = \text{SEM}$$

If the population is quasinormal, i.e., if the curve is not grossly skewed or bimodal, and if $N > 20$ or 30 in each group, then this method of constructing a skinny curve should be reasonably accurate and can be used for NHST. In practice, if the distribution is clearly non-normal in shape, a nonparametric statistical test is probably better to use than a t-test regardless of the number of datapoints.

What if My Sample Size Is Small, Say 8—10 per Group Rather Than >20—30?

This is a loaded question! In the real world, the t-test will not be very accurate for small data sets. The proper procedure is to use some kind of nonparametric statistical test instead (Chapter 12). However, given that so many published experiments use only 8—10 datapoints per group, and almost invariably use t-tests in this situation, some further discussion is in order.

If the sample distributions look quasinormal, can one still use the t-test out of convenience? In full disclosure, I have even done this myself. For example, in Table 9.1, I compare 10 sets of articles on the topic of that disease that were retrieved by a search engine versus 10 sets of articles related to biological processes, in terms of certain scored features. The P-value for the first test is .35, the second test .158, both well above .05; whereas in the third test, the P-value is very low, 1.07×10^{-5}, or 5000 times lower than .05.

The most appropriate statistical test to use in this situation is really the nonparametric Mann—Whitney U test (see Chapter 12), which gives P-values for the first test of .522, second test .347, and third test 4.4×10^{-4}, or roughly 100 times lower than .05. Notice that the t-test is more powerful than the U test for this data set and assigns too MUCH significance (gives lower P-values) compared with the U test.

Nevertheless, when the P-value given by the t-test is well above .05, the U test will generally also be above .05, and when the t-test gives a value that is extremely small, the U test will generally be extremely small too. It is important to note that Table 9.1 is not actually assessing the findings that my colleagues and I were reporting in the study, but merely characterizing whether the baseline characteristics of the article sets under study were similar or not. So it can be OK to use a t-test sometimes when the number of datapoints per group is only 8—10, as long as it is not important to obtain an accurate P-value!

TABLE 9.1   Example of a Series of t-Tests Performed on Two Groups That Have Only 10 Datapoints per Group

Query	#Articles First Time Slice	#Mesh Pairs	Mesh Pairs/Article
DISEASES			
Acute myocardial infarction	33,839	326,747	9.6559
Alcoholism	31,004	408,042	13.1609
AD	36,807	463,576	12.5948
Autism	7,464	104,235	13.9650
Colon cancer	46,470	767,217	16.5099
Cystic fibrosis	19,395	364,871	18.8126
Lupus	29,735	404,202	13.5935
Multiple sclerosis	23,374	364,889	15.6109
Schistosomiasis	7,863	132,983	16.9125
Suicide	28,543	410,629	14.3863
Mean	**26,449.4**	**374,739.1**	**14.5202**
SD	12,309.69	182,364.79	2.58
BIOLOGICAL PROCESSES			
Alternative splicing	15,616	468,767	30.0184
Apoptosis[ti]	38,518	883,333	22.9330
Bacterial evolution	16,140	363,602	22.5280
Endocytosis	31,775	819,690	25.7967
Hyperpolarization	7,534	214,117	28.4201
Ion transport	31,391	887,410	28.2696
Meiosis	14,067	295,162	20.9826
Microtubules	16,072	388,830	24.1930
Protein aggregation	31,932	851,583	26.6686
Working memory	11,792	188,437	15.9801
Mean	**21,483.7**	**536,093.1**	**24.5790**
SD	10,745.55	291,073.45	4.19
t-test *P*-value	*0.3495*	*0.1580*	***0.0000107***
U test *P*-value	*0.5222*	*0.3472*	***0.00044***

P-values are based on unpaired, two-tailed tests; the t-test was carried out assuming that the two groups do not necessarily have equal variance. See text for details. The *P*-values from the parametric t-test versus nonparametric Mann–Whitney U test are italicized.

Reprinted from Peng Y, Bonifield G, Smalheiser NR. Gaps within the biomedical literature: Initial characterization and assessment of strategies for discovery. Front. Res. Metr. Anal 2017;2:3. With permission.

C. STATISTICS (WITHOUT MUCH MATH!)

CHOOSING T-TEST PARAMETERS

To carry out a t-test using statistical software, you provide two sets of datapoints (e.g., highlighting data from two columns in an Excel spreadsheet) and set three parameters, which we will discuss in turn.

Paired or unpaired? If the datapoints can be paired up from one group to the other, then in most cases, it is desirable to perform the paired t-test. Pairing generally reduces variability within the groups and so it is generally more powerful than the unpaired test.

Tip: The obvious types of pairing involve paired experimental designs, for example, when comparing the same individuals before versus after some treatment is applied. However, as discussed in Chapter 5, remember that one can also pair up control versus experimental datapoints that were prepared or assayed in parallel on the same day, control versus experimental samples that were part of the same experimental run, control versus experimental patients seen by the same physician, etc.

One tailed or two tailed? The two-tailed t-test measures a difference between the control and experimental group, which may occur in either direction. This is appropriate when you have no prior expectations, and you are simply assessing the probability that sampling variability will produce changes between groups of a certain magnitude (or greater), which might occur in either direction.

Tip: If you do have prior indications that the experimental group will differ from the control group in one direction—this may be based on previous studies, pilot data, or even just based on your hypothesis—then it is better to pose a one-tailed test. For the same data, the one-tailed test is more powerful than the two-tailed test.

Equal variance or unequal variance between groups? As mentioned above, NHST assumes that the experimental group has the same variance (i.e., the same SD values) as the control group. However, if the groups differ substantially in their variances, then the t-test needs to use an alternative formula to take that into account.

Tip: Personally, I *always* set the t-test to the option for unequal variance, because this alternative formula will give correct results regardless of whether the variance is the same or different between groups.

A FINAL WORD

Null Hypothesis statistical testing (NHST) has a central position in teaching students how to do science. And, to be fair, NHST is an essential step in making sure that any effects that one observes cannot be explained simply by chance. Yet its misuse in practice has caused so many problems that we will continue to devote much of the next chapter to discussing them.

Tip: Remember that if you learn how to ask good questions, design experiments with an eye for controls and confounds, and conduct a study as if it were a court trial, then you will do good science—whether or not you choose exactly the right statistical test, and whether or not you obtain the most accurate *P*-value.

References

[1] Perezgonzalez JD. The meaning of significance in data testing. Front Psychol 2015;6:1293. http://dx.doi.org/10.3389/fpsyg.2015.01293.

[2] Gigerenzer G. Mindless statistics. J Socio-Economics November 30, 2004;33(5):587–606.

[3] Peng Y, Bonifield G, Smalheiser NR. Gaps within the biomedical literature: Initial characterization and assessment of strategies for discovery. Front. Res. Metr. Anal 2017;2:3.

10

The "New Statistics" and Bayesian Inference

STATISTICAL SIGNIFICANCE IS NOT SCIENTIFIC SIGNIFICANCE

An investigator carries out NHST to make sure that an observed difference is not likely to have occurred by chance. What is wrong with that?

For one thing, in real life, no two groups are never exactly identical; that is, the null hypothesis is ALWAYS wrong if you look hard enough! If I choose 100 corn plants at random from a cornfield, and then choose another 100 plants at random again, these two groups will not exhibit identical distributions of heights but will have slightly different means. Random sampling does not remove differences entirely. Thus if you have enough datapoints in each group, NHST will judge even very small differences to be statistically significant. For 100 corn plants, suppose a difference in mean height of 1 inch is at the border of achieving significance (i.e., rejecting the null hypothesis); then for 400 corn plants, a difference of 0.5 inch will suffice, and for 10,000 corn plants per group, a difference of just 0.1 inch will suffice. Such small differences may be statistically significant yet lack biological or ecological relevance.

The main fault with NHST, however, is that it has become used as the primary gatekeeper for determining if an experimental finding should be reported in a publication. This conflates at least four erroneous beliefs:

1. If two distributions are observed to be significantly different, then that finding must mean something important scientifically.
2. Statistical testing per se provides an important piece of evidence that the finding is valid.
3. If you observe an effect, even incidentally, during the course of an experiment, you are entitled to report it, nay, you *should* report it!
4. Negative results (those that do not achieve significance) are not worth reporting.

The result is that the scientific literature is filled with articles reporting effects that were found incidentally by trawling experiments fishing for differences that achieve significance. These articles neither feel the need to adduce independent confirmatory evidence that the effects are reliable and robust, nor do they systematically attempt to rule out sources of bias,

Data Literacy
http://dx.doi.org/10.1016/B978-0-12-811306-6.00010-5

confounds, and alternative explanations and investigate the underlying mechanisms that produce the effects. *In short, there is a close tie-in between the uncritical use of NHST and the use of naïve experimental designs* that produce experiments that lack meaning and reproducibility.

THE MAGICAL VALUE $P = .05$

By far the most common threshold for statistical significance assigns the risk of a false-positive finding at $P = .05$, meaning that an observed effect of that size might arise simply by sampling the same population 5% of the time. That is acceptable if you employ NHST for its original purpose, that is, as one of multiple cross-checks on the finding's validity.

However, when observing an effect that satisfies $P = .05$ becomes the criterion for publishing a finding, the risk that a reported finding will be false can rise dramatically, up to 50% or more! [1–6]. As discussed in Chapter 8, one reason is because investigators generally perform many statistical tests when mining experimental data. Suppose each test has a 5% chance of being false—then for every 20 tests that are performed, on average, one of those tests will achieve significance even if there are no true differences at all. The situation is even worse when the experimental design does not prespecify the outcomes to be examined, so that the investigators are free to create and compare subgroups against each other in a plethora of ways (i.e., data fishing).

Another problem with assigning too much importance to P-values is that they are highly nonlinear and unstable across replications. If I do an experiment and I see an effect size of 2.3, and if it is a robust effect, then repeating the experiment might produce an effect size of 2.4 or 2.5, i.e., not too greatly different. However, if 2.3 is at the threshold of significance at $P = .05$, it is possible that the effect 2.4 may produce a P-value of .005, and effect of 2.5 may have a P-value of .0005! **Small changes in effect size can produce wildly different P-values.** And, if two experiments have different sample sizes or sample variances, they could observe exactly the *same* effect size yet have entirely different P-values—one might be highly significant and the other does not achieve significance at all. Never forget that a small P-value does NOT imply a large effect size. Replicability means getting (about) the same effect size each time, NOT getting the same P-value each time!

HOW TO MOVE BEYOND NULL HYPOTHESIS STATISTICAL TESTING?

A variety of critiques have recommended a variety of solutions. David Colquhoun [1] has proposed that many of the flaws of NHST can be averted by insisting on a P-value $= .001$ as the threshold for significance instead of $P = .05$.

The movement that goes by the name of "the new statistics" [6,7] has urged scientists to use NHST in its proper context. This means the following:

1. reporting the actual P-value obtained, not simply saying "$P \leq .05$";
2. reporting the effect sizes and 95% confidence intervals, not just the P-values;

3. carrying out prospective power estimation and using it to determine the number of datapoints in each group (see Chapter 6);
4. avoiding pitfalls in data analysis such as data fishing and HARKing (see Chapter 8);
5. reporting experimental design, methods, and results fully (see Chapter 14).

I think that is good advice. Others have proposed replacing NHST entirely with alternative statistical frameworks such as permutation testing (see Chapter 12) or Bayesian inference methods (see below). In 2015, the journal *Basic and Applied Social Psychology* completely banned the reporting of *P*-values, t-tests, F-values, statements about statistical significance, etc. Yikes!

CONDITIONAL PROBABILITIES

NHST lacks the concept of conditional probabilities (Chapter 3), and scientists who think solely in terms of NHST will find it impossible to tackle problems such as the Prosecutor's Fallacy or the Monte Hall Problem. Moreover, conditional probabilities are central to Bayesian inference, which will be discussed in turn.

The Prosecutor's Fallacy

The Prosecutor's Fallacy shows the dangers of not taking conditional probabilities into account. All too often, in the real world, people have been prosecuted for crimes and the evidence for their guilt has been overstated by prosecutors due to improper understanding of the underlying probabilities. A particularly clear, fictional version of this is reprinted here, modified from http://www.dcscience.net/2016/03/22/statistics-and-the-law-the-prosecutors-fallacy/with permission:

> A murder has been committed on an island, cut off from the outside world, on which 1001 inhabitants remain. The forensic evidence at the scene consists of a measurement, x, on a "crime trace" characteristic, which can be assumed to come from the criminal. It might, for example, be a bit of the DNA sequence from the crime scene.

> Say, for the sake of example, that the probability of a random member of the population having characteristic x is $P = .004$ (i.e., 0.4%), so the probability that a random member of the population does *not* have the characteristic is $1 - P = .996$. The mainland police arrive and arrest a random islander, Jack. It is found that Jack matches the crime trace. There is no other relevant evidence. How should this match evidence be used to assess the claim that Jack is the murderer? (For illustration, we have taken $N = 1000$, $P = .004$.)

> Prosecuting counsel, arguing according to his favorite fallacy, asserts that the probability that Jack is guilty is $1 - P$, or .996, and that this proves guilt "beyond a reasonable doubt." The probability that Jack would show characteristic x if he was not guilty would be 0.4% i.e., P(Jack has x | not guilty) = .004. Therefore the probability of the evidence, given that Jack is guilty, P(Jack has x | Jack is guilty), is one $1 - .004 = .996$.

> But P(evidence | guilty) is not what we want. What we need is the probability that Jack is guilty, given the evidence, or in other words, P(Jack is guilty | Jack has characteristic x). *To mistake the latter for the former is the prosecutor's fallacy*, or the error of the transposed conditional.

Defence counterargument: Counsel for the defense points out that while the guilty party must have characteristic x, he is not the only person on the island to have this characteristic. Among the remaining N = 1000 innocent islanders, 0.4% have characteristic x, so the number who have it will be NP = 1000 × 0.004 = 4. Hence the total number of islanders that have this characteristic must be 1 + NP = 5. The match evidence means that Jack must be one of these 5 people, but does not otherwise distinguish him from any of the other members of it. Since just one of these is guilty, the probability that this is Jack is thus 1/5, or 0.2—very far from being "beyond all reasonable doubt."

The Monty Hall Problem

Another, more amusing example of conditional probability thinking is the Monty Hall Problem, named after the host of the game show *Let's Make a Deal*. A contestant is given the choice of three doors: behind one door is a car; behind the others, goats. The contestant chooses one door—say, number 1—and Monty (who knows what is behind the doors) opens door number 3, which has a goat behind it (Fig. 10.1). If you are the contestant, should you stick with the door you picked, number 1, or choose the other one (number 2)?

Without conditional probability thinking, the naïve answer is that the car has an equal probability 1/3 of being behind any door. Since the car might equally be behind number 1, number 2, or number 3, there is an equal chance of being behind door 1 or door 2 and it does not matter whether you stick or switch doors.

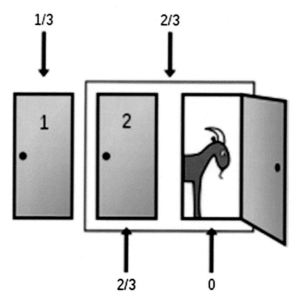

FIGURE 10.1   **The Monty Hall problem.** *Reprinted from https://commons.wikimedia.org/wiki/File:Monty_open_door_chances.svg with permission.*

With conditional probability thinking, the relevant question is: what is the probability of the car being behind door 1, GIVEN THAT Monty opened door 3? Monty did not do so at random! The probability of the car being behind door 1 is indeed 1/3, i.e., **P(door 1) = 1/3**; this implies that the probability that it is behind *either* door 2 or 3 = **P(door 2 OR 3) = 2/3**, since the probabilities must add up to 1 (Fig. 10.1). Monty's opening of one door gives us a crucial piece of information: **P(door 3) = 0**. Since we just said that P(door 2 OR 3) = 2/3, this implies that **P(door 2) = 2/3**. To recap, the probability that the car is behind door 1 is 1/3, and the probability that it is behind door 2 = 2/3. You better switch!

BAYES' RULE

Remember in Chapter 3 we promised to derive Bayes' rule? Let us say that P(A) is the probability that the Cubs will win a game tonight, and P(B) is the probability that it will rain tonight.

If the two events are independent of each other, then P(A and B) = P(A) × P(B).

If they are not independent of each other, it is possible that the probability of the Cubs winning will differ according to whether it rains or not. (It is less likely that the probability of rain will depend on whether the Cubs win or not, unless rain is God's tears and God is a Cubs fan.)

How do we express the joint probability of P(A and B) when A and B are not independent? P(A and B) will equal the probability of the Cubs winning, P(A), times the probability that it rains GIVEN THAT the Cubs win, which can be expressed as P(B|A).

That is:

$$P(A \text{ and } B) = P(A) \times P(B|A)$$

Equally, P(A and B) will equal the probability of rain, P(B), times the probability that the Cubs win GIVEN THAT it rains, which can be expressed as P(A|B).

$$P(A \text{ and } B) = P(B) \times P(A|B)$$

Putting these two equations together:

$$P(A \text{ and } B) = P(A) \times P(B|A) = P(B) \times P(A|B)$$

And so:

$$P(A) \times P(B|A) = P(B) \times P(A|B)$$

This identity is incredibly simple but powerful, since dividing both sides by P(B) gives:

P(A|B) = P(A) × P(B|A) / P(B)

which is **Bayes' rule**.

In words,

1. the **probability of the Cubs winning tonight, given that it rains tonight**, is equal to
2. the **probability of the Cubs winning (in general)**,
3. times the **likelihood that it rains during a Cubs win**,
4. divided by the **probability that it will rain tonight**.

Let us go through these in further detail.

1. The **probability of the Cubs winning, given that it rains**, is what we want to estimate. That is referred to in Bayesian terminology as the **posterior probability**. (The reason for calling it "posterior" is because we calculate it AFTER, not prior to, examining experimental data.)
2. The **probability of the Cubs winning (in general)** is called the **prior probability** in Bayesian terms. This probability may draw from objective data (e.g., what is the Cubs' win−lose record this season?) but no two investigators may choose the same way of estimating the prior probability, so this parameter remains more or less subjective. One person might only consider the win−lose record for the past month to be relevant, whereas another might include the win−lose record over the past 10 years. Still others might factor in scouting reports or recent injuries that might modify the odds of a win happening tonight.
3. The **likelihood that it rains during a Cubs win** is called the **likelihood** in Bayesian terms. This parameter is calculated by tabulating the percentage of games in which it rained and the Cubs won. Again, there is some uncertainty in how much observational data is relevant—on the one hand, enough examples are needed so that the proportion can be estimated reliably, but on the other hand, if one goes too far back into the past, the relation between rain and Cubs win may not be the same as holds tonight (e.g., because Cubs wins in the past may have involved different key players or different venues).
4. The **probability that it will rain tonight** is called the **marginal likelihood** in Bayesian lingo. One simply looks up the weather forecast. (The reason for calling it "marginal" is that it reflects two likelihoods: the likelihood that it rains during a Cubs win, and the likelihood that it rains during a Cubs loss.)

BAYESIAN INFERENCE

Bayesian inference is an alternative to NHST which applies Bayes' rule and conditional probabilities to experimental data. Let us use the baseball example just given to trace how a Bayesian would formulate the problem:

The **probability of the Cubs winning, given that it rains**, is the outcome that we want to estimate. That is referred to in Bayesian terminology as the **posterior probability**. In more complex situations we might want to estimate the distribution of posterior probabilities for all possible outcomes. For example, instead of estimating the chance of a Cub win, we might want to estimate the probability of the Cubs winning by each specific number 0, 1, 2, 3,… of runs. Often the goal of a Bayesian inference is simply to estimate the most likely outcome (among all possible outcomes), which means estimating the mean of the posterior probability distribution, rather than trying to estimate the entire distribution.

The **probability of the Cubs winning (in general)** is called the **prior probability**, or just the **prior**. In an experiment, **this represents the investigator's belief in the probability of the outcome**. The belief may be based on previous studies, preliminary pilot studies, predictions of theory, or subjective judgment. Sometimes the prior is given not as a single number but in terms of a probability distribution of its own; for example, one might estimate the chances of a Cub win as having a mean value of 0.35 but assume that the true value of the prior may vary from this (following, say, a normal distribution around the mean value).

The **probability that it rains during a Cubs win** is called the **likelihood**. In Bayesian inference, experimental data provide the observations that allow the likelihood to be calculated. The experimental data consist of a list of dates, whether it rained, and whether the Cubs won or not (and by how much). This allows one to assign a value of likelihood for every possible outcome. For example, how often did it rain when the Cubs won by 1, 2, 3,… runs?

Finally, the **probability that it will rain tonight** is called the **marginal likelihood**. Note that this parameter is a constant that is neither affected by the experiment nor by the outcome. The shape of the posterior probability distribution, and the estimated mean of that distribution, does not depend on the value of the marginal likelihood. Therefore, when the goal is to estimate the most likely outcome (among all possible outcomes), the marginal likelihood does not need to be calculated explicitly, but can be thought of as a constant or a normalizing factor.

Let us put this all together: The main idea of Bayesian inference is that the investigator starts with an imperfect guess, the prior, and carries out an experiment (which estimates the likelihood). **Plugging both prior and likelihood into Bayes' rule results in a much improved estimation of which outcomes are most likely, over either parameter alone.** A common exercise is to run the Bayes' rule equation separately for all possible values of the outcome, to see which outcome is predicted to be the most likely overall. That outcome is the one which is most compatible with both the observed experimental data and the estimate of the prior.

The uncertainty in calculating the prior is Bayesian inference's theoretical Achilles' heel, since sometimes there are no objective data for estimating the prior, and one must use gut feelings or even assume that all outcomes are equally likely. However, sometimes there is strong evidence for a particular value of the prior, whereas the experimental data are weak or inconclusive. In such cases, the Bayesian inference provides much better results than null hypothesis testing alone, which only considers the experimental data. Running an experiment, itself, provides objective evidence that can improve the investigator's estimate of the prior, and so one can adjust the prior and rerun the equations for even better performance.

An Example of Bayesian Inference

Baseball and Bayes seem to go well together. Let us consider a baseball player, Rod, who has hit three times in his last three games, for a total of four times up to bat. In other words, his batting average equaled 0.75 for his last three games. What is the chance that he will have a batting average of 0.75 over his next three games? We set up Bayes' rule:

$$P(A|B) = P(A) \times P(B|A) / P(B)$$

P(A|B) = the probability of hitting a batting average of 0.75 over the next three games, given that his average was 0.75 for the past three games.

P(A) = the probability of hitting a batting average of 0.75 in general.

P(B|A) = the likelihood that his batting average was 0.75 for the past three games, given that he obtains a batting average of 0.75 over the next three games.

P(B) = the probability of hitting a batting average of 0.75 over the past three games.

As Bayesians, the first thing we do is estimate P(A), the prior probability of hitting .75 in the first place. Long-term batting averages in the major leagues are mostly over 0.2 and less than 0.4 (Fig. 10.2), so let us estimate the prior as having a mean value of 0.3, following a normal distribution with standard deviation of 0.05. Then, a batting average of 0.4 is 2 SD above the mean, and a value of 0.75 resides a full 9 SD above the mean! The probability of hitting .75 or greater consistently is small indeed; let us charitably call it roughly one in a million, or **.000001**.

Next, we examine the experimental data (his performance over the past three games) to estimate the likelihood. Ideally, it would be nice if we had a sliding window over a long period of time, to see his batting average during one set of three games related to the subsequent set of three games. However, all that is available is the past three games, in which he hit three out of four runs. Let us assume, for illustration's sake, that batting averages for a player tend to hover around a mean value and follow a normal distribution, then the experimental data can be estimated as a normal distribution centered around a mean value of 0.75.

We now have two normal curves. One is given by the prior, which is centered around 0.3 and is based on a huge amount of long-term data averaged over many players. The other is given by the experimental data, which covers one player averaged over just three games and

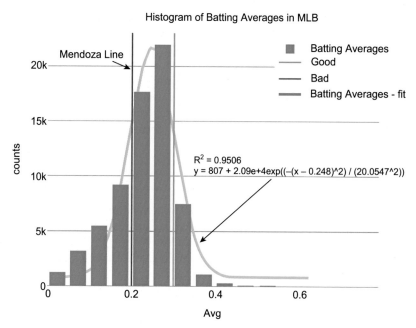

FIGURE 10.2   Batting averages in Major League Baseball. *Reprinted from https://plot.ly/~mkcor/305/histogram-of-batting-averages-in-mlb/ with permission.*

C. STATISTICS (WITHOUT MUCH MATH!)

is centered around 0.75. The experimental data are not very definitive, robust, or reliable, yet they cannot be ignored!

The likelihood **P(B|A)** is actually pretty high. If you told me that Rod hit 0.75 in his next three games, then the chance that he hit 0.75 in the last three games is pretty good—let us say **0.5**.

Finally, the marginal likelihood, that is, the probability of hitting a batting average of 0.75 over the past three games, is not quite as low as the prior, since temporary fluctuations do occur. One might look up how often a three-game fluctuation of this size or greater has occurred among major league players, but here we simply will assign the probability as **0.02**, or happening once in 50 games on average.

Putting the equation together, **P(A|B)** = (**0.000001** × **0.5**)/**0.02** = .000025.

That is, the chance of hitting 0.75 in the next three games is only 25 in a million, even though he hit 0.75 in the last three games. The very low prior probability accounts for this. We could construct the entire posterior probability distribution by asking what is the probability associated with each possible outcome—a batting average over the next three games of 0.1, 0.2, 0.3, and so on, by estimating the parameters of the Bayes equation for each possible outcome. The most likely outcome will not be centered at 0.3, as in the prior, but will be shifted to some extent toward 0.75 to reflect the influence of the experimental data.

Calculating Bayes parameters, and solving the posterior probability distribution, can be very difficult in general (e.g., when the parameters do not follow normal distributions). It may not be feasible to solve the equations analytically (i.e., by plugging parameters into the equations and solving algebraically). Instead, the estimations may need to be done by sampling randomly across many different combinations of parameters, using methods known by names such as Markov chain Monte Carlo simulation or Gibbs sampling. Until recently, the massive computation involved was a major reason for the limited popularity of Bayesian inference in analyzing experimental data. However, modern computers can run open source software packages in R or Python programming languages to carry out these estimations. So although those who use Excel and even statistical software may find Bayesian inference to be largely out of reach, the climate is rapidly changing as scientists are becoming more data- and computer literate. A Bayesian open source version of statistical software, JASP, is becoming available for free download at https://jasp-stats.org/. This is still in the early stages (the most recent version is release 0.8), but worth tracking since availability of easy-to-use software may be the limiting factor for popular acceptance of Bayesian methods.

COMPARING NULL HYPOTHESIS STATISTICAL TESTING AND BAYESIAN INFERENCE

Null hypothesis statistical testing is sometimes referred to as being based in a "frequentist" view of statistics, which has certain inherent philosophical problems. For example, NHST assumes that each possible outcome has an underlying "true" probability that is fixed, although unknown. At the same time, the experimental datapoints, which are acquired by sampling from the population, are viewed as uncertain and randomly changing from experiment to experiment. Yet as scientists, we have a strong tendency to believe what we directly observe—that is, to take the data as fixed and given—and regard the distribution of possible outcomes as uncertain values to be estimated. Thus, a case can be made for flipping the assumptions of NHST on their heads.

Most importantly, null hypothesis statistical testing does not take into consideration conditional probabilities. In particular, it does not incorporate the **a priori probability** that a specific finding will occur. Not every finding is equally expected or plausible! If I say that Madame Oculo can ascertain the outcome of coin flips performed 1000 miles away by gazing into a crystal ball, you would be very skeptical about accepting that finding, even if she called 7 coin flips in a row correctly (the odds of being correct being 1 in 128, or $P = .0078$). Even if she did this twice ($P = .000061$), you would probably still resist concluding that she has supernatural skills and look for other explanations.

To give another example, if I find that protein X binds to protein Y significantly better than to a negative control, is that surprising or not? NHST is of no help here; **P-values and statistical significance are not a measure of how surprising a finding is**. Rather, the answer depends on what is known about the structures of proteins X and Y, the known mechanisms by which they could bind, and how promiscuous proteins X and Y are (that is, how many other proteins are they known to bind?).

In NHST one constructs 95% confidence intervals around the mean, whereas in Bayesian statistics, one constructs **95% credible intervals** instead. Depending on the value of the prior and the relative strength of evidence behind the prior versus the observed experimental data, the 95% credible interval may not correspond to the 95% confidence interval, and in particular, may not be symmetric around the mean.

SYSTEMATIC REVIEWS AND METAANALYSES

Birds got to sing, scientists got to write review articles. And probably for the same reason! The vast majority of review articles are really opinion pieces, in which the author chooses certain topics to discuss and selects articles that are particularly seminal, influential, important, controversial, or recent. The author may review an entire field or focus on only a few articles, to highlight, compare, or criticize them. Such **narrative reviews** often synthesize findings into generally accepted facts, identify questions which are unsettled, and propose promising new research directions on the horizon.

In contrast, a new type of review called **systematic reviews** is becoming increasingly popular as a way to analyze evidence across multiple studies. Systematic reviews follow guidelines that attempt to maximize rigor and minimize subjective biases:

1. Systematic reviews are designed prospectively. That is, the authors define at the outset, as precisely as possible, the research question and its exact scope to be reviewed. They define the types and sources of evidence that will be considered relevant, and the exact search strategies to be used to find the evidence (online, in libraries, and/or through personal communications). Many systematic review designs are published as articles in their own right as protocols, prior to undertaking the review.
2. Systematic reviews are comprehensive (within scope). Great effort is made to find all relevant evidence. Note that different reviewers may define their scope somewhat differently. For example, some reviewers are only interested in studies published in English, whereas others are interested in studies published in any language, but only within the past 15 years. Published studies may be searched exhaustively (even hand searching the

table of contents of certain journals that are key in that field), and in some (but not all) cases, attempts may be made to find unpublished "gray" literature such as material presented at conferences or posted online.

3. Systematic reviews attempt to assess the quality and bias of the available evidence. They take into account the study sample size, randomization schemes, and the extent and types of potential biases when evaluating the studies. Many systematic reviews also try to ascertain how much evidence may be missing due to studies that have been completed yet not published. This so-called "file drawer effect" (Chapter 6) happens most often when studies produce negative or uninterpretable results, or when sponsors of the research do not like the outcomes.

Systematic reviews are most prevalent in the field of evidence-based medicine, where reviewers assess clinically relevant questions such as whether a drug or class of drugs are effective in treating a particular disease, whether they are likely to have side effects, whether a cancer screening test performs well, etc. In clinical medicine, large, randomized, controlled trials are considered the most reliable type of evidence followed by nonrandomized trials, clinical studies, and case reports. Systematic reviews are less common but increasingly being carried out in other biomedical and social sciences—for example, studies of how genetic alterations are related to disease, neuroimaging of human brain in people performing certain tasks, effectiveness of humanitarian interventions, and so on.

Metaanalyses perform statistical analysis of data that is combined across multiple studies. Generally, these analyze summary data (such as effect sizes and variances) that are presented in published articles. Metaanalysis is potentially a very powerful way to identify findings that are reproducible and achieve consensus, as well as to identify findings that fail to be generally supported. As shown in Fig. 10.3, it is possible to consider five different studies that examined the same phenomenon. Even if none of the studies individually achieved statistical significance, when taken together, it is possible to discern strong statistical support for an effect.

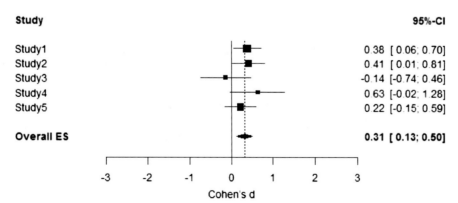

FIGURE 10.3 **A forest plot showing effect sizes observed in five independent studies.** None of the studies are individually significantly different from an effect size of 0. Yet the combined metaanalysis shows an overall effect size that is moderate (Cohen's d, or SD units = 0.31) and significantly different from 0. *Reprinted from http://daniellakens. blogspot.com/2014/08/on-reproducibility-of-meta-analyses.html.*

C. STATISTICS (WITHOUT MUCH MATH!)

Metaanalyses are not free from their own challenges, since (not surprisingly) how reviewers choose to include and weight evidence across different studies can affect the conclusions. A new and welcome trend in metaanalysis is to obtain and analyze the raw data across studies, e.g., data from individual patients, rather than just using the sanitized data which is tabulated in the published articles.

Not all metaanalyses are systematic reviews, and not all systematic reviews are metaanalyses. However, both types of research examine a set of related findings in multiple contexts and across multiple studies.

References

[1] Colquhoun D. An investigation of the false discovery rate and the misinterpretation of *p*-values. R Soc Open Sci 2014;1:140216. http://dx.doi.org/10.1098/rsos.140216.

[2] Simmons JP, Nelson LD, Simonsohn U. False-positive psychology: undisclosed flexibility in data collection and analysis allows presenting anything as significant. Psychol Sci November 2011;22(11):1359—66.

[3] Chavalarias D, Wallach JD, Li AH, Ioannidis JP. Evolution of reporting P values in the biomedical literature, 1990-2015. JAMA March 15, 2016;315(11):1141—8. http://dx.doi.org/10.1001/jama.2016.1952.

[4] Hupé JM. Statistical inferences under the null hypothesis: common mistakes and pitfalls in neuroimaging studies. Front Neurosci February 19, 2015;9:18. http://dx.doi.org/10.3389/fnins.2015.00018.

[5] Greenland S, Senn SJ, Rothman KJ, Carlin JB, Poole C, Goodman SN, Altman DG. Statistical tests, P values, confidence intervals, and power: a guide to misinterpretations. Eur J Epidemiol April 2016;31(4):337—50. http://dx.doi.org/10.1007/s10654-016-0149-3.

[6] Cumming G. The new statistics: why and how. Psychol Sci January 2014;25(1):7—29. http://dx.doi.org/10.1177/0956797613504966.

[7] Cumming G, Calin-Jageman RJ. Introduction to the new statistics: estimation, open science, and beyond. New York: Routledge; 2017.

ANOVA

ANALYSIS OF VARIANCE (ANOVA)

The idea behind the ANOVA is very simple. In words: Tabulate all of the variability in your experiment and divide into **variability ACROSS groups versus variability WITHIN groups.** If the variability ACROSS groups is much greater than the variability WITHIN groups, then at least one of the groups has a mean that is significantly different from the others.

ONE-WAY ANOVA (ONE FACTOR OR ONE TREATMENT)

"One way" means that there is a single outcome measure that is being compared across two or more groups simultaneously. Let us walk through the math for one-way ANOVA for an experiment that involves a total of k groups, each one having n datapoints in each group.

The variability of a group is simply its **variance**, which we know (from Chapter 3) is the **Sum of Squares** of the datapoints in the group divided by **df**, its **degrees of freedom**. If there are a total of n items x_i in group 1, and the mean of the datapoints is called m_1, then

For group 1, Sum of Squares $= SS_1 = (x_1 - m_1)^2 + (x_2 - m_1)^2 + (x_3 - m_1)^2 + \ldots (x_n - m_1)^2$

Within-group SS added up across all groups $= SS_w = SS_1 + SS_2 + SS_3 + \ldots SS_n$

Within-group variance $= SS_w / df_w$

where $df_w = nk - k$.

(Note that some textbooks call the variance the "mean square" when talking about ANOVA, and they refer to within-group variability as the "error." This terminology is unfortunate because it obscures the fact that ANOVA uses SS and df concepts, just as normal curves and correlations do.)

In contrast, the across-group variance looks at the deviation of each group's mean from the overall grand mean M of all datapoints in the experiment.

Across-group SS added up across all groups $= SS_a = (m_1 - M)^2 + (m_2 - M)^2 + (m_3 - M)^2 + \ldots (m_k - M)^2$

Across-group variance $= SS_a / df_a$

where, $df_a = k - 1$.

Data Literacy
http://dx.doi.org/10.1016/B978-0-12-811306-6.00011-7

Finally, we compute the ratio of across-group variability versus within-group variability:

$$\text{Across-group Var/Within-group Var} = (SS_a/df_a) \, / \, (SS_k/df_k)$$

If this value is much greater than 1, then at least one of the groups is significantly different from the others. A ratio of variances follows the **F-distribution** (Chapter 3). To calculate the threshold for significance, one looks up a table of F-values for a given value of df in the numerator (across-group Var, $df = k - 1$) and df in the denominator (within-group Var, $df = nk - k$), and for the desired type I error (usually 0.05). See Box 11.1 for an example worked out by hand.

Tip: When the one-way ANOVA is used to compare two groups, the test is mathematically equivalent to the t-test and the F-value is equal to the square of the t-value, or $F = t^2$. Thus, if a t-value of ~ 2 is at the threshold of significance at $P = .05$, then this corresponds to F-value of ~ 4.

Because across-group Var + within-group Var = Total Var, sometimes people will not compute the across-group variance directly, but instead will compute total variance and subtract the within-group variance, as follows:

$$\textbf{Total Variance} = SS_{tot} = (x_1 - M)^2 + (x_2 - M)^2 + (x_3 - M)^2 + \dots (x_n - M)^2$$
$$df_{tot} = nk - 1$$
$$\text{Total Var} = SS_{tot}/df_{tot}$$
$$\text{Across-group Var} = \text{Total Var} - \text{Within-group Var.}$$

ANOVA IS A PARAMETRIC TEST

Like the t-test, the ANOVA is a parametric test, meaning that it makes certain assumptions about the distributions of its datapoints:

- each group is sampled from a normal distribution,
- each datapoint in the same group is independent of the others, randomly sampled, and follows the same underlying population distribution;
- the variance of each group is similar.

Just as we said that the t-test can be applied to data which are quasinormal and can be applied when the variances are "not too different," so can the ANOVA be used to give relatively robust results when these assumptions are violated.

Tip: Never use an ANOVA when there are less than 5 observations or datapoints per group, and a minimum of at least 20 per group is generally preferred.

There are specialized statistical tests that can be used to measure how badly an experimental data set strays in terms of normality and equal variance although these are most useful when the number of datapoints per group is relatively large ($> \sim 30$). (Equal variance goes by the infelicitous term **homoscedasticity**, which is in keeping with the idiosyncratic terminology used to describe ANOVAs!) Simply plotting and visualizing the data are the best ways to assess normality and equal variance, to decide if ANOVA is a suitable test. If not, nonparametric tests can be done instead (see Chapter 12).

BOX 11.1

AN EXAMPLE OF A ONE-WAY ANALYSIS OF VARIANCE TEST WORKED OUT BY HAND

Shown is a toy data set, with $n = 8$ samples per group, $k = 2$ groups, $Nk = 16$ total samples in the study.

Group 1	Group 2
1	1.5
2	4
3.5	3.5
2	3
8	4
3	2.5
4	1
2.5	3

Total variance:

Grand mean = M = sum of all x_i /N = 3.03125

$SS = $ sum $(x_1 - 3.03125)^2$
$= (1 - 3.03125)^2 + (2 - 3.03125)^2 +$
$(3.5 - 3.03125)^2 + (2 - 3.03125)^2 +$
$(8 - 3.03125)^2 + (3 - 3.03125)^2 +$
$(4 - 3.03125)^2 + (2.5 - 3.03125)^2 +$
$(1.5 - 3.03125)^2 + (4 - 3.03125)^2 +$
$(3.5 - 3.03125)^2 + (3 - 3.03125)^2 +$
$(4 - 3.03125)^2 + (2.5 - 3.03125)^2 +$
$(1 - 3.03125)^2 + (3 - 3.03125)^2$
$= 4.12 + 1.06 + 0.2197 + 1.06 + 24.688$
$+ 0.00097 + 0.969 + 0.282 + 2.345 +$
$0.969 + 0.2197 + 0.00097 + 0.969 +$
$0.2822 + 4.12 + 0.00097$
$= 41.307$
$Df = Nk - 1 = 16 - 1 = 15$
Total variance = $SS/df = 2.7538$

Within-group variance:

Group 1 mean = $26.5/8 = 3.25$
$SS = (1 - 3.25)^2 + (2 - 3.25)^2 +$
$(3.5 - 3.25)^2 + (2 - 3.25)^2 + (8 - 3.25)^2$
$+ (3 - 3.25)^2 + (4 - 3.25)^2 +$
$(2.5 - 3.25)^2$
$= 5.06 + 1.5625 + 0.0625 + 1.5625 +$
$22.5625 + 0.0625 + 0.5625 + 0.5625$
$= 31.9975$
$Df = nk - k = 16 - 2 = 14$, Var1 = $31.9975/14 = 2.2855$
Group 2 mean = $22.5/8 = 2.8125$
$SS = (1.5 - 2.8125)^2 + (4 - 2.8125)^2 +$
$(3.5 - 2.8125)^2 + (3 - 2.8125)^2 +$
$(4 - 2.8125)^2 + (2.5 - 2.8125)^2 +$
$(1 - 2.8125)^2 + (3 - 2.8125)^2$
$= 1.722 + 1.41 + 0.473 + 0.0351 + 1.41$
$+ 0.098 + 3.29 + 0.0351$
$= 8.4732$
$Df = nk - k = 16 - 2 = 14$, Var2 $= SS/df = 0.605$
Var1 + Var2 = 2.8905

Across-group variance:

- Grand mean = $3.03,125$, mean1 = 3.25, mean2 = 2.8125
- $SS = (3.03125 - 3.25)^2 + (3.03125 - 2.8125)^2 = 0.0478 + 0.0478 = 0.0957$
- **$Df = k - 1 = 1$**, Var = $SS/df = 0.0957$

Now, test significance:

- **$Var_{across}/Var_{within} = 0.0957/2.8905 \ll 1$**
- Since the ratio is less than one, we immediately know that this will not be significant!
- The ratio will follow the F-distribution with numerator $k - 1 = 1$, denominator $nk - k = 14$ degrees of freedom
- Use an F-distribution table to look up the P-value.
- Or...let your statistical software do the work for you!

TYPES OF ANOVAs

When samples can be paired across groups, this reduces the within-group variability, and this increases the power of the ANOVA to detect small differences as being significant.

To see how this occurs, consider the following two situations:

In the first, we examine systolic blood pressure measured in two independent groups of subjects. One group (subjects 1–3) is measured under baseline conditions, and the other (subjects 4–6) is exposed to a drug that might alter blood pressure. Note that the subjects in the no-drug condition will vary among themselves according to their baseline blood pressures, but the subjects given drug will vary BOTH because they have different baseline blood pressures (not explicitly measured here) AND ALSO because they will vary in their response to the drug.

	Subject 1	Subject 2	Subject 3
No drug	100	110	120
	Subject 4	Subject 5	Subject 6
Given drug	140	100	130

In contrast, if the same subjects are measured before versus after drug treatment, the values are paired across the two groups:

	Subject 1	Subject 2	Subject 3
Before drug	100	110	120
After drug	110	120	135

Here, subjects 1–3 still vary in their baseline blood pressures, but the drug-treated group will ONLY vary in how each subject responds to the drug. The variability within the drug-treated group will be less with paired than with nonpaired designs.

This pairing is a simple example of **repeated-measures** ANOVA. The same subject might be measured at many different times, e.g., weekly, resulting in many paired measurements per subject.

Another related way to reduce within-group variability is **blocking**:

In an unblocked one-way ANOVA, all datapoints in a group are considered together:

	Group 1	Group 2
No drug	100, 110, 120	
Given drug		140, 100, 130

In a blocked **two-way** ANOVA, different subgroups, say males and females, are assessed separately

	Group 1		Group 2	
	Males	Females	Males	Females
No drug	100	110, 120		
Given drug			100	140, 130

Blocking reduces overall within-group variability and hence increases the power of the ANOVA, because when computing the Sum of Squares, males are only compared against males and females against females.

A **factorial** ANOVA is similar to a two-way or blocked ANOVA, but it explicitly looks for interactions among the factors when the experiment employs factorial designs (Chapter 5). In the example just given, the ANOVA can detect whether subjects respond differently to the drug, as well as whether different genders respond differently to the drug.

In a **three-way** ANOVA, one outcome is related to three different independent variables. **Multivariate** ANOVAs are performed when there are two or more outcome variables measured for each subject or datapoint. And this is only a partial list—seems that every type of experimental design has a corresponding ANOVA designed to handle it. Generally, you will be using statistical software to perform ANOVA and simply need to choose which type of ANOVA is desired, as well as the desired threshold level of significance (type I error).

THE ANOVA SHOWS SIGNIFICANCE; WHAT NEXT?

Obtaining a significant result in an ANOVA test, say at $P = .05$ or better, merely says that SOME groups or factors are different than others, but does not pinpoint WHICH groups are different. So, after all that effort, you still need to carry out multiple comparisons anyway! Usually these comparisons are done pairwise among each of the groups or subgroups in the experiment. But there is a subtle difference between carrying out pairwise comparisons alone versus doing so after performing ANOVA. Namely, in the former case, each comparison has a false-positive rate of (say) 5% and there is no assurance that any comparisons should be significant. In contrast, the ANOVA provides evidence that a significant effect does exist, with an overall risk of a false positive of 5% (or less, depending on the F-value that was obtained).

So, given a positive ANOVA test, one does further testing using t-tests to identify which groups or factors are different from the others. It is necessary to correct post hoc comparisons for the number of t-tests that are performed, to obtain a more realistic estimate of the false-positive rate.

CORRECTION FOR MULTIPLE TESTING

How to correct for multiple testing most appropriately is an entire subject in itself. Corrections are not specific for ANOVA tests but need to be applied whenever multiple statistical tests are carried out.

Colquhoun's Correction

Perhaps the simplest procedure is to focus only on tests that achieve a P-value of .001 or less. This takes care of most sins that might have been committed in experimental design as well as multiple testing and has the virtue of restricting the investigator's attention to findings that are large and most likely to be reproducible. However, this is a very **conservative** procedure, that is, it reduces the power of each test by bending over backward to avoid false positives. Most scientifically meaningful findings will not satisfy this criterion.

Bonferroni Correction

The most famous method of correcting for multiple tests is the Bonferroni procedure. If the desired type 1 error for each test is 0.05, and you carry out 10 tests, then the corrected threshold for significance is $P = .05/10 = .005$. That is, none of the 10 tests will be deemed significant unless they achieve a P-value of .005 or less. For 100 tests, the corrected threshold is $P = .05/100 = .0005$. This is, again, a very conservative procedure.

The underlying assumption of the Bonferroni procedure is that each test is independent of the others, which is often *not* the case at all. For example, suppose I am comparing two authors, A and B, to see how frequently they use pronouns (say, "he" vs. "she") in the body of their published books. This results in two tests: author A versus author B for "he," and author A versus author B for "she." But is the frequency of the word "he" in a person's writings independent of the word "she"? Maybe or maybe not. If mentions of "he" and "she" are correlated across one or both person's writings (see Chapter 13), then the Bonferroni correction (here, correcting for only two tests) will actually overcorrect and result in a P-value threshold that is too stringent, i.e., too low. In sociology, suppose I am studying the elderly and testing whether two subgroups tend to apply for retirement benefits at different ages. Test A compares low-income versus high-income people, and test B compares people who identify as belonging to one race or another. If there is a correlation between race and income, these two tests are not independent, and again, the Bonferroni procedure will be too conservative.

Other Correction Procedures

A variety of correction methods have been proposed and popularly applied for post hoc analysis of ANOVA test results. These go by names such as the Newman–Keuls method, Tukey's honest significance test, Scheffé's method, and others. They differ in their details and how conservative they are, and none is clearly preferred for all experiments. Choosing one of these goes beyond the scope of this book. With statistical software, it is easy to examine and compare how these different methods will alter the P-values of a finding in your own experiment.

Benjamini–Hochberg Procedure

This is very popular in fields such as bioinformatics where many thousands of statistical tests may be carried out simultaneously. The idea is to specify a desired False Discovery Rate (FDR), which is **the fraction of positive tests that are false positives**. The Benjamini–Hochberg method at a specified FDR of 5% identifies the threshold level of significance, such that only 5% of the tests that achieve that P-value or less will be false positives. This generally gives a corrected *P*-value threshold that is less conservative than applying the Bonferroni correction. (Bonferroni adjusts $P = .05$ to a corrected threshold of $.05/n$, where n is the total number of statistical tests carried out, whereas the Benjamini–Hochberg procedure adjusts the threshold to $i \times 0.05/n$, where i is the number of positive tests.)

Permutation Testing

Permutation testing is a nonparametric method of carrying out statistical testing, which has a built-in correction for multiple testing. Permutation tests are discussed in detail in Chapter 12.

You should be aware that if you carry out 20 different t-tests during the course of analyzing your data, on average, one of the tests will achieve significance at $P = .05$ just by chance. So it is important to compare significance levels before versus after correction (by any one of these methods) to judge for yourself whether to take the finding seriously or not. Just as we said there is no magical value of $P = .05$ that determines whether a finding is truly significant or not, there is no magical value of correction that determines whether the finding is worth studying and reporting, either.

12

Nonparametric Tests

INTRODUCTION

My elder son used to work at Abercrombie and Fitch, so I went there one day to see if I could buy some clothes. I am an average-sized guy, yet I felt like a giant when looking at their selection! Their target customer is slim and tall, a demographic that is a shrinking minority in a land where two-thirds of adults are overweight or frankly obese.

I feel the same way when I look at t-tests and ANOVAs and try to fit my experimental data to their requirements. These parametric tests are aimed at a narrow target demographic of data that fit a normal distribution and have large sample sizes—unrealistic for the vast majority of healthy experiments in the real world! So, do not think of nonparametric statistical tests as exotic or advanced methods but quite the reverse. Nonparametric methods are designed for real data: skewed, lumpy, having a few warts, outliers, and gaps scattered about.

Why, then, are parametric tests so much more popular and widely taught than nonparametric tests? Abercrombie and Fitch deliberately aimed their clothes at "cool kids," and certainly data that fit a normal distribution have "cool" mathematical properties that make them easy to teach and use. For example, if you are studying a population whose features lie on a normal curve, then the distribution of samples will follow a t distribution and the sampling distribution or "skinny curve" can be employed for statistical testing even at low sample sizes (<20 datapoints). A more important reason for the popularity of parametric tests is that they are generally more powerful than nonparametric tests. This is a valid reason for favoring parametric tests for well-behaved (quasinormal) data sets.

Unfortunately, when parametric tests are utilized on data sets that are not normal, they are still more powerful than nonparametric tests—as we saw in Chapter 9, they can overestimate the true level of significance, which is more robustly and more accurately estimated by nonparametric tests. Unlike t-test and ANOVA, nonparametric tests:

- do not make assumptions about data that are often wrong or hard to check,
- are valid even with small sample sizes and with ordinal measures,
- do not need complicated correction factors for when the groups do not have equal variance.

Tip: In practice, when analyzing an experiment *I always check my primary outcomes using both parametric and nonparametric statistics.* **If the two types of analyses give similar values, this can be used as justification for reporting the results of the parametric tests. Conversely, if parametric versus nonparametric tests give substantially different significance values, this provides evidence that nonparametric statistics are more appropriate in this situation.**

Although I will be showing you how to compute the nonparametric tests by hand, most statistics software can carry out nonparametric tests as readily as doing a t-test. Although Excel does not provide nonparametric tests, at least at present, there are both free and commercial add-on software kits available that can allow you to perform nonparametric tests (and a variety of other statistical tests such as ANOVA) within the Excel environment.

THE SIGN TEST

The simplest nonparametric test, the **sign test**, is also the least powerful. It is based on the statistics of coin flips. For example, in the single sample sign test, the hypothesis being tested is that the median of the population being sampled is some prespecified number M. If M is indeed the median value, half of the sample points should be above M and half below M. But due to sampling variability, it is possible that the number of datapoints above and below M might not be exactly the same even if the population median is M. How different can they be before we should conclude that the true median is not M? Let us say there are 20 datapoints in the sample; 5 are less than M and 15 are greater than M. The probability of that happening by chance is the same as the probability of getting 5 heads out of 20 flips. The two-tailed P-value $= .0414$, suggesting that the true median of this sample is probably different from M.

The two sample sign test follows the same logic but is applied to comparing two paired groups. For example, consider test scores in eight math students measured before versus after eating a candy bar:

Student	A	B	C	D	E	F	G	H
Before:	72	45	67	83	83	49	23	77
After:	54	67	34	72	90	49	34	54
Sign:	−	+	−	−	+	Tie	+	−

The chance of observing 3 +'s out of 7 total (note that we ignore any ties if present) are the same as observing 3 heads out of 7 flips or a two-tailed P-value of .705. This suggests that the candy bar had no significant effect on test scores (at least, none that we can detect with confidence).

The sign test is not very powerful, because it only considers the direction of change but not the magnitude. On the other hand, it is OK to perform the sign test even when there are very few datapoints in the sample, even as few as five or less.

THE WILCOXON SIGNED-RANK TEST

This paired test is more powerful than the sign test because it considers both the direction and ranks of differences across groups. For example, consider these two paired groups and compute the following measures:

Group 1:	12	45	67	83	13	9
Group 2:	5	67	34	12	90	12
Paired difference	+7	−12	+33	+71	−77	−3
Rank pairs by abs value	−3	+7	−12	+33	+71	−77
Ranks	1	2	3	4	5	6

(Ignore pairs that are equal, i.e., group differences $= 0$. However, if nonzero difference values are ties, these are averaged and given the same rank, e.g., if first three difference values were ties, the ranks would be given as 2, 2, 2, 4, 5, 6.)

We compute two summary statistics: W+ is the sum of ranks for those pairs that showed positive differences $2 + 4 + 5 = 11$ (based on 7, 33, 71), and W− is the sum of those pairs that showed negative differences $1 + 3 + 6 = 10$ based on ($-3, -12, -77$).

Based on the values of W+ and W−, use a look-up table to calculate the level of significance (these can be found online). (Note that the number of datapoints N in the look-up table refers to the number of paired datapoints excluding ties, not the total number across both groups; in the example above, there are six paired datapoints.) For a two-tailed test, use the lesser of W+ and W− in the look-up table. For a one-tailed comparison expecting that group 1 is less than group 2, use W+ (if significant, W+ should be less than the threshold of significance shown in the look-up table). For a one-tailed comparison expecting that group 2 is less than group 1, use W−.

In the case shown, where W− $= 10$ and N $= 6$, the difference is not significant (indeed, with only six paired datapoints, all of the pairs would need to have greater values in one group for the difference to be significant at $P = .05$).

THE MANN–WHITNEY U TEST

This is a workhorse among nonparametric tests, because it applies generally to comparing two unpaired groups. It assumes that each datapoint within a group is sampled independently from the same underlying distribution. However, unlike the t-test, the U test can be carried out on very small sample sizes; ordinal measures are permitted (e.g., movie reviews rated from 1 to 5 stars) as well as interval and ratio measures. And of course, the U test does not assume that the data distribution is normal or even quasinormal.

To perform the U test (see Box 12.1), rank all the observations across both groups, beginning with 1 for the smallest value, but keep track of which group each datapoint comes from. If some rank values are tied, assign them a rank equal to the midpoint of unadjusted rankings (that is, if the data set is 4, 6, 6, 9, then the ranks are 1, 2.5, 2.5, 4).

BOX 12.1

AN EXAMPLE OF THE U TEST WORKED OUT BY HAND

Consider the same two groups we compared above, but consider them now as unpaired groups.

Group 1:   12   45   67   83   13   9
Group 2:   5   67   34   12   90   12

Order them: *5, 9, 12, 12, 12, 13, 34, 45, 67, 67, 83, 90*

Ranks with ties: *1, 2, 4, 4, 4, 5, 6, 7, 8.5, 8.5, 9, 10*

R1 = 2 + 4 + 5 + 7 + 8.5 + 9 = 35.5
R2 = 1 + 4 + 4 + 6 + 8.5 + 10 = 33.5

$$U1 = N1N2 + (N1)\cdot(N1 + 1)/2 - R1$$
$$= 36 + (6)\cdot(7)/2 - 35.5 = 21.5$$

$$U2 = N1N2 + (N2)\cdot(N2 + 1)/2 - R2$$
$$= 36 + (6)\cdot(7)/2 - 33.5 = 23.5$$

Look up U1 in a look-up table: for $N1 = 6$, $N2 = 6$, alpha = 0.05, two-tailed, the critical value of U1 is 5 and for alpha = 0.01, the critical value of U1 is 2. Note lower values of U are more significant. Here, the U scores are far above the critical values, so the differences are nowhere near significance.

Next, take the N1 datapoints that came from sample 1 and add up all their ranks = R1. Also add up the ranks for the N2 datapoints in sample 2 = R2.

U is then given by:

$$U1 = N1N2 + (N1)\cdot(N1 + 1)/2 - R1$$
$$U2 = N1N2 + (N2)\cdot(N2 + 1)/2 - R2$$

Take U1 or U2, whichever is smaller; use look-up tables to calculate the significance level. (A smaller value of U is more significant.)

Tip: Although the U test is often referred to as testing whether two groups have the same median, it is really a global test of difference of the two data distributions. That is, differences in variance, skew or other shape parameters can also affect the U test.

How can one report the effect size, that is, how big of a difference exists between the two groups, in a nonparametric format? If the two distributions have similar shapes, one can simply give the difference in medians in group 1 versus group 2. Or, more generally, one can consider all of the paired datapoints between each value of group 1 and each value of group 2, and state what percentage of pairs exhibit higher values in group 1.

EXACT TESTS

These are called "exact" because one examines a particular scenario, in which there are a finite, countable number of possible outcomes. Count up all of the ways that all outcomes can be produced and then count up how many of these produce the outcome that was actually observed.

For example, if you flip a coin 10 times, there are $2^{10} = 1024$ possible outcomes. The first flip can be heads or tails, the second flip heads or tails, and so on. Suppose you actually get 8 heads. What is the probability that you will get 8 (or more) heads out of 10 flips?

Note that the number of ways of choosing k items from n things $= n!/k! \cdot (n - k)!$
where $n! = n \times (n - 1) \times (n - 2) \times (n - 3)\ldots \times 2 \times 1$.

P(0 heads, i.e., all tails) $= 1/1024$
P(1 head) $= 10/1024$
P(2 heads) $= 45/1024$
P(3 heads) $= 120/1024$
P(4 heads) $= 210/1024$
P(5 heads) $= 252/1024$
P(6 heads) $= 210/1024$
P(7 heads) $= 120/1024$
P(8 heads) $= 45/1024$
P(9 heads) $= 10/1024$
P(10 heads) $= 1/1024$

Then, P(8 or more heads) $=$ P(8 heads) $+$ P(9 heads) $+$ P(10 heads) $= (45 + 10 + 1)/1024 = .0547$.

Fisher's Exact Test

This exact test is applied when asking whether an observed outcome differs between two groups, when the outcomes are expressed as proportions. That is, given group 1 having N1 datapoints and group 2 having N2 datapoints, we observe a particular outcome happens A times in group 1 and B times in group 2, and ask whether the proportion A/N1 is significantly different from B/N2.

This problem can be conceptualized as a 2×2 contingency table, where the groups are in rows and the outcomes are in columns. The method counts up all the ways that you could achieve different outcomes, given the observed row and column totals, and assign probabilities to each. The null hypothesis is that groups (rows) are independent of columns (outcomes).

For example, suppose a market research firm is trying to identify commercials, which are effective in inducing viewers to buy their product (say, a bacon-flavored beer). They hold viewing sessions for two groups and then follow up to see whether they have bought the beer or not within a day of viewing.

	Bought Beer	Did Not Buy Beer
Commercial 1	30	10
Commercial 2	10	30

Are the two commercials significantly different in their effectiveness? I used an online Fisher's Exact Test calculator (http://www.langsrud.com/stat/fisher.htm) to compute this and found that the two-tailed *P*-value is .000305, which is highly significant.

The parametric version of this test is called the chi-square test of independence. I do not cover the chi-square test in this book, because Fisher's exact test gives an exact value of probability, whereas the chi-square test only gives approximate values and has the baggage associated with other parametric tests (assumes random sampling, normality, large numbers of datapoints per group, etc.).

NONPARAMETRIC T-TESTS

The Mann–Whitney U test is the true nonparametric counterpart of the t-test and gives the most accurate estimates of significance, especially when sample sizes are small and/or when the data do not approximate a normal distribution.

However, there is something familiar and comforting about using t-tests! When one has a large sample size ($N \gg 30$) but the data are skewed, it is worth examining log- or square root-transformed values of the data to see if they become more quasinormal (see Chapter 7). If the data pass a test for normality (included in most statistical software), it is then OK to perform a t-test using the transformed datapoints.

Another alternative when N is large is to convert the datapoints to their ranked values (i.e., rank 1 is the smallest value, rank 2 is the next smallest, and so on) and carry out a regular t-test on the rank-transformed datapoints. As always, it is advisable to set the t-test parameters for unequal variance across groups.

NONPARAMETRIC ANOVAS

Nonparametric versions of ANOVA tests are common enough that they have their own names. The nonparametric one-way ANOVA is called the Kruskal–Wallis test, and the nonparametric repeated measures ANOVA is called the Friedman test (named after the economist Milton Friedman, who invented it). These tests (performed on rank-transformed data) do not require that the data distributions are normal, but they do assume that datapoints are independent of each other and that each group has roughly equal variance. Rather than assessing differences of means across groups, these tests assess differences in median values, and they have their own look-up tables (not the F distribution).

PERMUTATION TESTS

Permutation is just a fancy word for randomly shuffling the datapoints in an experiment (Fig. 12.1). But permutation is more than a technique—it is a basic way of thinking about experiments. I am a big fan of permutation testing, both because it is so conceptually simple and powerful, and because it can be applied so widely. Permutation tests are nonparametric exact

FIGURE 12.1   Shuffling a deck of cards is a familiar way to permute their order randomly.

tests, but there are advantages to using them even in situations where the sample sizes are large and the data are normally distributed.

To give an example of how permutation was employed in a study from my own laboratory, my colleague Vetle Torvik and I once sought to identify sites by which microRNAs could potentially bind to messenger RNAs (mRNAs). At the time, the rules by which microRNAs bind to mRNAs were unknown, so we took a strictly statistical approach [1]. The set of then-known microRNA sequences (roughly 22 nucleotides long) were scanned for their extent of complementarity against a reference set of several thousand mRNA sequences (each of which might be several thousand nucleotides long), and we tabulated how many "hits" (complementary binding interactions) we got that were of length 10, 11, 12, … up to perfect complementarity (22 nucleotides). But how to calculate what distribution would be expected by chance?

What we did was to randomly shuffle the nucleotide sequences of each microRNA and repeat the process, counting the hits of all shuffled microRNAs on the set of all mRNAs. And we did not randomly shuffle just once, but multiple times, so that we could accurately estimate the 95% confidence interval of the number of expected hits of length 10, number of expected hits of length 11, and so on. In this case, 10 sets of shuffled sequences were adequate to discern significant trends.

We found that the number of hits associated with microRNAs began to exceed the hits produced by randomly shuffled sequences, and the size of the difference increased as the hit length got progressively larger (Fig. 12.2). The same method could be used to discern some of the rules that determined microRNA targeting. For example, there was statistical evidence that several distinct microRNAs tend to bind near each other (Fig. 12.3) and that microRNAs tend to target multiple mRNAs (Fig. 12.4).

C. STATISTICS (WITHOUT MUCH MATH!)

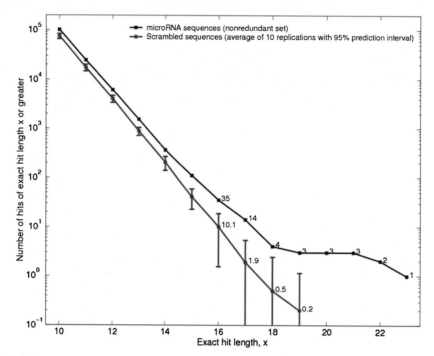

FIGURE 12.2 **microRNAs and their scrambled counterparts interact differently with the population of human mRNAs [1].** Shown are all exact hits ≥10 bases long produced on human mRNAs by the set of microRNAs versus the average of 10 replications of scrambled control sequences. Shown is the number of hits as a function of exact hit length. Only the longest hit was counted: e.g., for a hit of length 18, the two subsets of length 17 in the same hit position were not counted.

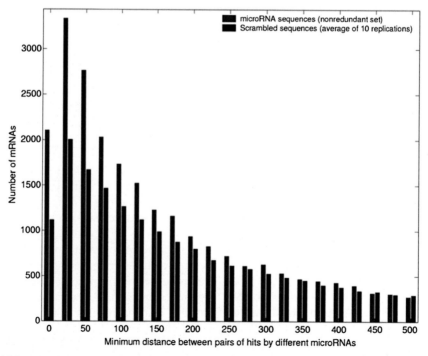

FIGURE 12.3 **Number of distinct mRNA sequences that received hits from two or more distinct microRNAs, as a function of the minimum distance between hits [1].** Distance of 0 or 1 was excluded because this might be produced by partial overlap of microRNA sequences.

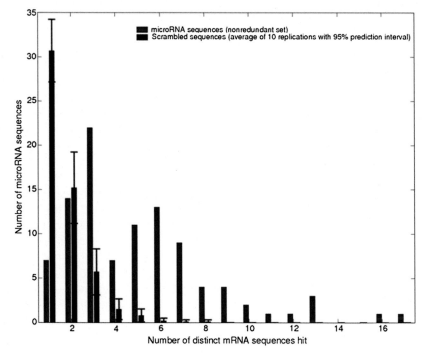

FIGURE 12.4   Individual microRNAs hit multiple targets on the candidate list, more often than expected by chance [1].

We wrote up the results and submitted the manuscript, and to our surprise, the reviewers criticized us for shuffling the nucleotides randomly. As it happens, a biological sequence not only has a particular nucleotide composition (proportions of A, C, G, and T) but also a characteristic dinucleotide composition as well (proportions of AA, AG, AC, AT, CG, CT, etc.). So, we repeated the entire experiment, this time shuffling the nucleotide sequences in blocks of two. Fortunately, the results still held!

How Permutation Tests Work

As the example showed, the general framework is to compare your observed experimental data set with multiple sets of appropriately shuffled data to see if the observed outcome could have been produced simply by chance. You examine the outcomes in the shuffled data and see if the observed outcome is estimated to occur in <5% of the shuffled sets. If so, your observation is significantly different from chance at $P < .05$.

Data might be shuffled in several ways. For example, recall the experiment described earlier in this chapter, where two groups of 40 people each were exposed to different commercials and were tested to see if they bought beer or not afterward.

	Bought Beer	Did Not Buy Beer
Commercial 1	30	10
Commercial 2	10	30

Instead of using Fisher's exact test to estimate the significance of the finding, we could have carried out permutation testing: To do this, we pool the entire data set of 80 people from the experiment and divide the pool in half randomly, creating two new groups of 40. We tabulate how many people in each group bought beer in the original experiment. Then, we divide the 80 people into two groups randomly again, and tabulate how many bought beer. We do this again and again, let us say 100 times, and ask: In how many of the shuffled data sets did at least 30 people in one group buy beer? If it occurred in 1 of the shuffled data sets, the P-value is estimated as 1/100 or .01. (This P-value is estimated, not exact, because if you had run a different set of 100 shuffled groups, you might have observed fewer or more than 1 data set in which at least 30 people in one group bought beer.)

How many replications of the data sets are needed? This depends on the size of the observed difference and the precision to which you want to calculate the P-value. Sometimes 10 replications are enough (Figs. 12.2–12.4), and sometimes thousands of replications may be necessary. The massive computation needed for permutation testing is no longer a practical limitation using modern computers and available open source software packages programmed in R, Python, and other languages.

In fact, the major limitation in using permutation testing is not computation power, but the question of how best to shuffle the data sets appropriately. I mentioned that when we were shuffling nucleotide sequences, we did not initially realize that pairs of adjacent nucleotides (dinucleotides) have biological meaning, so that randomly shuffling an entire sequence versus shuffling blocks of dinucleotides would produce different baselines. In general, you need to ask whether the datapoints within the same group are truly independent of each other or have interactions that need to be maintained during shuffling.

Using Permutations to Correct for Multiple Testing

A different use for permutation testing is to correct for the effects of multiple statistical tests, and this can be applied even when utilizing t-tests or other parametric tests. For example, suppose I am measuring the expression levels of 1000 different genes in liver samples taken from subjects in two groups. I want to carry out t-tests for each gene separately, but the single-test P-value of .05 (i.e., type I error of 5%) is not appropriate; instead, I need to correct this P-value for the fact that I am doing 1000 t-tests.

To find the correct P-value threshold for significance in this situation, I can pool all subjects into one pool and divide them randomly into two groups. I do this, say, 500 times. Each time, I carry out the 1000 t-tests between the two shuffled groups, rank them from smallest to largest P-values and record the P-value threshold that represents the lowest 5% of tests. By plotting each of the P-value thresholds obtained across all the shuffled data sets, I obtain a distribution of P-value thresholds; the corrected P-value threshold (for a type I error of 5%) is the one that shows that value or smaller in only 5% of the shuffled data sets.

The permutation method is better than the Bonferroni correction method (Chapter 11) in this case, for two reasons. First, Bonferroni would automatically set the P-value threshold at $.05/1000 = .00005$, a very low, worse-case scenario that may or may not be optimal. Second, Bonferroni assumes that the different genes are independent of each other, which may not be the case here, since many genes are coregulated together under certain conditions that might apply in this experiment. Note that the permutation method shuffled subjects between groups, but we did *not* shuffle the genes themselves, so that any interactions among the genes themselves were preserved and would be reflected in a better adjusted P-value threshold.

Reference

[1] Smalheiser NR, Torvik VI. A population-based statistical approach identifies parameters characteristic of human microRNA-mRNA interactions. BMC Bioinf September 28, 2004;5:139.

13

Correlation and Other Concepts You Should Know

LINEAR CORRELATION AND LINEAR REGRESSION

In plain English, "correlation" means that two things are related and as one varies, so does the other. That is pretty close to the statistical meaning, too. However, there are a surprisingly large number of nuances to learn to use correlations in analyzing data.

The most common type of correlation is **linear correlation**, which is closely tied to the concept of **linear regression**. I will discuss the two here as a single topic. **Linear regression finds the best way to fit a straight line to a given set of datapoints. Linear correlation measures how well the best-fit line describes the relation between two variables.**

Let us consider a simple example. Heights vary in the population, and heights tend to show a relationship with age over the life span. What part of the variability in height is due to age?

Age is the independent variable x and treated as if it has no measurement error. Height is the dependent variable y. Correlation analysis makes certain key assumptions about this scenario, which are as follows:

- The dependent variable y (e.g., height) follows a normal distribution at each age x.
- The height of each individual is independent of other individuals.
- The variance of the height at each age is constant (does not vary with age).
- You can visualize the x,y pairs by plotting them as a scatter plot on a two-dimensional graph (Fig. 13.1).

Our goal is twofold. First, we identify, among all possible straight lines that run through the data set, the one that is the best fit to the observed data (Fig. 13.2).

All straight lines are of the form $y = a + bx + \varepsilon$ where y and x are the dependent and independent variables, respectively, a is the intercept (i.e., the point where the line intersects with the y-axis), b is the slope of the line, and ε is a randomly varying term, which accounts for the fact that y is not fully determined but varies according to a normal distribution. The parameter ε includes measurement error as well as ALL OTHER influences not due to x. The

169

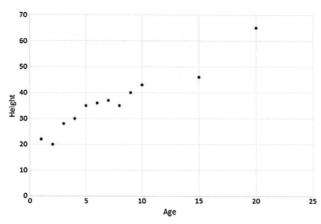

FIGURE 13.1   Idealized graph of height (in inches) versus age (in years) for 13 fictional individuals between ages 1 and 21.

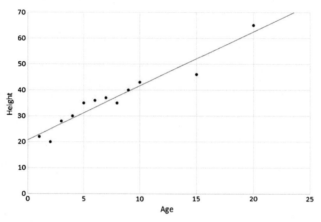

FIGURE 13.2   Graph of height versus age showing the best-fit line through the data set. Same data as Fig. 13.1 but with best-fit line added.

equation for the best-fit line gives the slope and intercept that most closely fits the overall data set.

Second, we measure how well the best-fit line approximates the data. This is done by measuring the vertical distance from each point to the line (Fig. 13.3), squaring that value, and adding these up over all points. In other words, we calculate the Sum of Squares of y calculated from the best-fit line (= SS_{line}).

Tip: Actually, the Sum of Squares calculation is used to identify the best-fit line in the first place: That is, among all possible straight lines through the data set, the best-fit line is the one that has the smallest Sum of Squares value.

C. STATISTICS (WITHOUT MUCH MATH!)

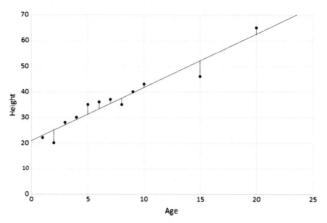

FIGURE 13.3 **Measuring the vertical distance from each point to the best-fit line.** *By Oleg Alexandrov—self-made with MATLAB, Public Domain, https://commons.wikimedia.org/w/index.php?curid=4099808.*

SS_{line} is then compared to the Total Sum of Squares of the data set, which is calculated relative to the mean value of y (= SS_{tot}). Graphically, SS_{tot} is the Sum of Squares calculated relative to the horizontal line through the mean value of y (Fig. 13.4).

The better the line fits the data set, the smaller SS_{line} will be relative to SS_{tot}. The r^2 parameter is defined as the proportion of total variability explained by the best-fit line, defined as

$$r^2 = (SS_{tot} - SS_{line})/SS_{tot}$$

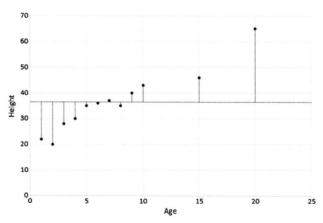

FIGURE 13.4 **A horizontal line through the mean value of y in a data set.** The distance from each point to the mean value is shown. *Reprinted from https://commons.wikimedia.org/wiki/File:Variables_independantes.svg.*

C. STATISTICS (WITHOUT MUCH MATH!)

The same relationship holds when formulated in terms of variances rather than Sum of Squares:

$$r^2 = (Var_{tot} - Var_{line})/Var_{tot}$$

(Note the similarity to the ANOVA—in both cases, total variance is divided into two parts, whose ratio is used to provide insight.)

Some important points are given as follows:

1. There is always a best-fit line to any data set, though the best-fit line may or may not be a very good fit. At best, all points will lie directly on the line, so that $r^2 = 1$. At worst, the best-fit line will be the horizontal line through the mean value of y so that $r^2 = 0$.
2. The process of fitting the best line to a data set is what is meant by **linear regression.** The slope b of the line shows you how strongly height is related to age—that is, for every increase in age of 1 year, height increases on average by b inches.
3. The best-fit line also allows you to predict the most likely value of height for any given value of age, even for ages that you have not observed yet. For example, for age 17, you can draw a vertical line through age 17 and see where it intersects the best-fit line; the height at that point is the predicted height for age 17. This process is called **interpolation** and assumes that the relationship between height and age is indeed linear, at least at the point where interpolation is performed.
4. The process of calculating r^2 is what is meant by **linear correlation.** The square root is called the **correlation coefficient r**, or more precisely, the **Pearson linear correlation coefficient**, and it varies between −1 and 1. A *negative* value for r means that an increase in x is associated with a *decrease* in y.

WHAT CORRELATIONS MEAN AND WHAT THEY DO NOT

Correlations are based on a linear model relating y to x. Of course, in most real-life data sets, that is a highly idealized approximation at best, and y may be related to x via complicated nonlinear relationships (e.g., parabolic function, hyperbolic tangent, or logistic function, to name a few). Just because a best-fit line exists does not mean that it tells you anything deep or meaningful about the data. Consider these following tips:

1. If $r = 0$, that does NOT mean that x and y are unrelated. **They could have a nonlinear relationship**.
2. If $r = 1$, this does NOT mean that x and y are related in any direct fashion. Both could be driven separately by a third or unknown factor Z! See Fig. 13.5.
3. Do not extend the best-fit line beyond the observed datapoints! This is called **extrapolation**. Best-fit lines have low reliability at both extremes of the data set.
4. Pearson linear correlation coefficient r values are highly influenced by outliers and are not reliable when n is very small. To see why, imagine a data set that has only two points. It will always fit perfectly on a line!
   Tip: In general, small data sets tend to have inflated estimates of r.

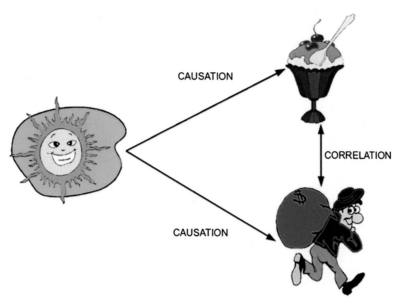

FIGURE 13.5 **Correlation is not causation!** Warm weather increases consumption of ice cream, and it increases the prevalence of crime. Ice cream and crime are correlated but not causally related. *By Rcragun—Own work, CC BY 3.0, https://commons.wikimedia.org/w/index.php?curid=7746472.*

5. The value of r does not tell you anything about the strength of the relationship between y and x. **The strength of the relationship is given by the** *slope* **of the best-fit line, not by the correlation coefficient**.

6. The value of r says how well changes in x account for changes in y. However, you may sometimes see calculations in scientific articles, in which the authors verify that a given value of r is significantly different from 0 and present a *P*-value to measure the extent of significance. Even very low values of r may be significantly different from 0, especially in large data sets, so do not be overly impressed if you see a low *P*-value.

7. Correlations can be asymmetrical. Suppose Frank and Marie have fixed salaries, which provide each of them with average baseline bank account balances, and they share an employer who occasionally gives both of them bonuses. However, there may not be any common force driving them below their baseline value. So, Frank and Marie's balances may be positively correlated for balance increases but uncorrelated for balance decreases.

8. Correlations are misleading if you do not measure the relation of y to x across the entire data set. For example, suppose we ask: Does height correlate with basketball performance? If this question is asked across the entire population, we find that (say) r = 0.9, close to 1. However, if we study only players in the National Basketball Association (NBA), we find, say, r = 0.001, which is essentially zero. Why? Because the general population ranges from ~3 to ~7 feet, but NBA players only range from ~6 to ~7 feet. They show relatively little variability in heights. As a result, the variability in their

performance across individuals is predominantly due to differences in other parameters (speed, reaction time, skill, motivation, etc.).

Tip: Any time you use the dependent variable to restrict the set of datapoints to be correlated, you fall victim to a dangerous statistical fallacy.

Another example of this fallacy is trying to correlate performance on entrance exams with subsequent performance in graduate school. Since the performance on those very exams helps determine who gets into school in the first place, the variability on exam scores will be relatively low among those who get accepted versus the whole population of those taking the exam, and so the correlation of their exam scores with later performance will appear to be low.

9. Finally, we have so far discussed correlations in cases where the independent variable, x, is a fixed and known value (e.g., age, publication date of an article, or longitude of a location). However, correlations can also involve independent variables that are subject to measurement error or are estimated rather than known. The math is different (e.g., different values for degrees of freedom are used), but the ideas are basically the same.

NONPARAMETRIC CORRELATION

The Pearson linear correlation coefficient r gives misleading or wrong values when outcome measures do not follow a normal distribution, when outliers are present, and when the relation between y and x is nonlinear. These pitfalls can be avoided by using the Spearman nonparametric correlation coefficient rho, which is calculated by substituting the numerical values of each x and y by their x rank values and y rank values (Fig. 13.6). That

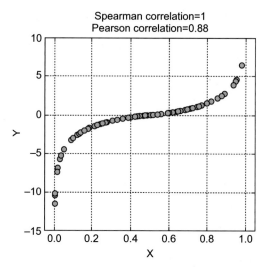

FIGURE 13.6 **Data set showing a nonlinear but steadily increasing relationship between x and y, such that the Spearman rho = 1 and the Pearson r = 0.88.** *By Skbkekas—Own work, CC BY-SA 3.0, https://commons.wikimedia.org/w/index.php?curid=8778554.*

is, plot each x and y point such that the x-axis shows the rank values of x and the y-axis shows the rank values of y. This allows you to see whether an increase in x leads to an increase (or decrease) in y, regardless of the magnitude of the change and regardless of whether or not the change occurs steadily (follows a straight line). Like r, rho ranges from -1 to 1. But unlike r, rho is not sensitive to outliers.

Tip: In practice, when analyzing correlations, I always compute r and rho values and compare them. If the values are very similar, that is evidence that the data are linearly related, whereas if the values are substantially different, that is a red flag that the data are not linear and the Spearman rho values should be used.

MULTIPLE LINEAR REGRESSION ANALYSIS

In principle, multiple linear regression is a simple extension of linear regression, but instead of relating one dependent outcome variable y to one independent variable x, one tries to explain the outcome value y as the weighted sum of influences from multiple independent variables x_1, x_2, x_3,....

In equation form,

$$y = k + ax_1 + bx_2 + cx_3 + ... + \varepsilon$$

where k is the intercept of the line on the y-axis; a, b, and c are the slopes of the relations between y and x_1, x_2, and x_3, respectively; and ε is again the random error term. This is basically just plotting the best-fit line, but instead of doing it on a two-dimensional graph, it is plotted through $n + 1$-dimensional space (n independent variables plus one dependent variable).

An example of multiple regression would be to relate height y not simply to age but to $x_1 = $ age and $x_2 = $ body weight. Knowing both age and weight should permit better estimation of a person's height than either parameter alone.

What makes multiple regression models so much harder to fit than linear regression is not the number of variables per se, but rather the fact that **the different independent variables may not be independent of each other!** For example, consider $x_2 = $ body weight versus another potential independent variable, $x_3 = $ body metabolic index (BMI), which measures body fat. These two measures are not redundant, and they do measure different things, but they are not entirely independent, either. A person could be very thin and very tall and so weigh a lot, but across a population of individuals, in most cases, a person's weight and their BMI are likely to be positively correlated. As well, if x_1 is age in years and x_4 is years since reaching puberty, these two variables are likely to be highly correlated with each other among a population of adults.

Independent variables need to be disentangled from each other mathematically, to optimize the multiple regression equation, also known as fitting the multiple regression model. A multiple regression model is fitted by throwing out independent variables that have no significant relation to y and by normalizing independent variables in a manner that removes their influence by other variables. (For example, if x_1 is age in years and x_4 is years since reaching puberty, we might replace x_4 in the regression equation by a new variable $x_5 = x_1 - x_4 =$ age *at* puberty, which is uncorrelated with x_1).

The final fitted model explains how an increase of one unit in each independent variable, say x_1, causes a change of a units in the outcome y, while holding all other x's constant. Of course, being a linear approximation, the equation may not fully or accurately account for the relationships between variables and outcome. As a rule, in practice, linear models are usually examined first to see how well they perform, before contemplating more complicated nonlinear regression methods.

There are also several other issues that make multiple regression more complicated than simple linear regression. One is the problem of **overfitting**, which means that fitting the equation perfectly for one data set might not generalize well to new observations. We will discuss ways to avoid overfitting later.

Another issue is how to encode independent variables that are not numerical. For example, height is not only related to age but also to gender. Can we write an equation that encodes gender in a meaningful numerical form? Yes, **dummy variables** serve that purpose. We add one new variable, say x_6, that reports male versus female status. One gender is assigned $x_6 = 0$ on this variable, and the other is assigned $x_6 = 1$ (does not matter which). This allows the equation to reflect different relationships among variables when the outcome is measured in a male versus a female.

Finally, always remember that multiple regression "explains" the effects of one variable on observed outcomes, but this does not mean that the explanatory variable has any direct effect on the outcome. Although knowing a person's weight may help predict his/her height, we would not be fooled into thinking that a person's weight causally *determines* his/her height. In regression, as in correlation, both the outcome and the correlated independent variable may simply be influenced by a common influence [1] (Fig. 13.5).

LOGISTIC REGRESSION

Logistic regression is a very popular type of multiple linear regression that can handle outcomes that are yes versus no rather than numerical values. For example, a regular multiple regression model might deal with age at death as an outcome—possible values being death at age 50, 63, 71, and so forth. In contrast, if we measure whether a person is alive or not 10 years after heart surgery, the outcome is either yes or no. Many outcomes in science and in life are yes or no—registered as an organ donor or not? Won the Nobel Prize or not? Arrested for speeding or not?

At first glance, a linear model is not suitable for studying yes/no outcomes. Recall the following basic equation:

$$y = k + ax_1 + bx_2 + cx_3 + \ldots + \varepsilon$$

The values on the right side can take any real number (positive or negative) and so the outcome y might be any real number too. Is it possible to make an equation such that the outcome is only 0 or 1 and still fit a linear model?

That is the beauty of the **logit transformation**, the trick that makes logistic regression work.

Step 1 is to change the goal of the equation slightly. Instead of predicting the outcome value directly, we estimate the *probability* that the outcome will be 1. But we cannot set up an equation that looks like the one that follows:

$$P(y) = k + ax_1 + bx_2 + cx_3 + \ldots + \varepsilon$$

Because P(y), as a probability, only ranges between 0 and 1, whereas as we said, the right side can equal any real number.

Step 2 is to create a new outcome variable that is fully determined by the probability P(y) but can range over any positive real number. This is the **odds ratio**, or the probability that the outcome will occur divided by the probability that it will not occur; that is, $P(y)/(1 - P(y))$. As P(y) ranges from 0 to 1, the odds ratio ranges from 0 to positive infinity. But we are still not there yet:

$$P(y)/(1 - P(y)) = k + ax_1 + bx_2 + cx_3 + \ldots + \varepsilon$$

This still does not work since the left side is positive whereas the right side may be either positive or negative. So we need to make one more transformation.

Step 3 is to take the log of the odds ratio, or the **log odds**, as the outcome to be modeled

$$\text{Log } [P(y)/(1 - P(y)] = k + ax_1 + bx_2 + cx_3 + \ldots + \varepsilon$$

this is the **logit transformation** and can be rewritten:

$$P(y)/(1 - P(y)) = e^{k + ax_1 + bx_2 + cx_3 + \ldots + \varepsilon}$$

And with a little algebra, we have finally:

$$P(y) = e^{k + ax_1 + bx_2 + cx_3 + \ldots + \varepsilon}/(1 + e^{k + ax_1 + bx_2 + cx_3 + \ldots + \varepsilon})$$

In this formulation, both the left side and right side of the equation can only range from 0 to 1, even though the parameters $k + ax_1 + bx_2 + cx_3 + \ldots + \varepsilon$ can be any real numbers!

Apart from the use of the logit transformation, logistic regression is basically just multiple linear regression, subject to the modeling aspects and caveats discussed in the previous section. The values of coefficients a, b, c,… have a slightly different meaning, of course, since you are modeling the log odds instead of predicting the value of y directly (Step 3, above). For example, a 1 unit increase in the independent variable x_2 will produce b units change in the log-odds ratio.

MACHINE LEARNING

At heart, machine learning refers to any mathematical or computer-based methods that seek to **identify meaningful patterns** within data sets [2]. These patterns may be used for a wide variety of purposes, ranging from classification (Is this defendant guilty or not?) to regression (What are the chances that this smoker will have a stroke within 5 years?) to the many sequential steps that are involved in guiding a self-driving car on the highway.

Supervised Learning

Machine-learning methods that employ **supervised learning** depend on having access to a body, training set, or **corpus**, of examples that are marked as "positive" or "negative." These are often marked manually by experts, in which case the examples are referred to as a **gold standard**. Sometimes the training data may be automatically generated; for example, the list of subscribers to a singles dating service may be assumed to consist largely of unmarried people, although there may be some "errors" in which married people signed up for the service. Use of automatic training data is referred to as **semi-supervised learning**.

The basic idea is that some type of mathematical or computer-based model is constructed using the training sets, and once it is fitted optimally for the training sets themselves, the model is tested to see how well it performs on new data.

The performance of a machine-learning model depends not only on having good training data but also on selecting good features (independent variables) and encoding/adjusting the features optimally. (We saw an example of adjusting independent variables when we discussed multiple linear regression.) The positive and negative training examples are used to fit, or train, a classification model so that it gives the highest performance on the training data.

However, to avoid **overfitting**, the model is not evaluated on exactly the same examples it was trained on. A common practice is to use **10-fold cross validation**, in which 90% of the training data are used to train the model, and then the model is tested on the remaining 10%; this is performed 10 times and an average is taken. This procedure still does not fully predict how well the model will perform when fed new test data that it has not seen before, especially for new data that appear in a new context. For example, a machine-learning model that recognizes proper names (such as Margaret Thatcher) in newspaper articles may not work as well when applied to different sources of text, such as hospital patient notes or Facebook posts.

Performance can also suffer when the training or testing data are highly **imbalanced**, that is, when most examples in the real world are negative and only relatively few are positive. (This is the issue of prior probability, as discussed in Chapters 3 and 10; even a very good model will produce mostly false positive predictions when the a priori probability is very low.)

Reinforcement Learning

This is a form of supervised learning, but instead of training the computer with all of the training examples in advance, the computer is given an unknown example one at a time and is then either reinforced positively (if the example is positive) or negatively (if the example is negative). This means that the more data the computer sees, the better it gets over time. And, this also implies that the computer continues to train its model as it sees new test examples in the real world, as long as it receives feedback on the correctness of its decisions. This allows it not only to continually correct itself but also to change the model itself if the new test examples begin to diverge substantially from the earlier examples.

Unsupervised Learning

The goal of unsupervised learning methods is to identify patterns based, not on any external reference set of examples, but on the relationships among the datapoints themselves. The most popular type of unsupervised learning is **clustering**, in which some measure, or **metric**, of similarity is defined between pairs of datapoints. Then, the most similar datapoints are clustered together into the same group.

For example, suppose we consider the data set of freshmen students living in a dormitory, and we want to group them according to their common interests. There is no absolute best way to do this and no absolute endpoint that tells us how many clusters there should be, nor how big the clusters should be. We might cluster together the students who are interested in sports versus the ones interested in academics, or we might separately cluster those interested in hockey versus those interested in soccer, or we might cluster the students instead according to where their parents live. Different solutions may be equally good depending on the purpose of clustering.

A simple and popular method of clustering is **K-means clustering**. The investigator simply chooses the final number k of clusters to be produced and employs an **algorithm**, that is, a sequential set of procedures, that will put any set of n items into k clusters. These methods are simple and often easy to compute. However, the clustering solution may not be unique— for example, the order in which the procedure looks at each item may affect the final clustering. The clustering may not be optimal for its intended purpose, and the choice of k itself may not be optimal. For example, if I clustered the set of freshmen students using k = 2, dividing them into two clusters may not satisfy my (or their) needs to put them together.

Another popular approach is **agglomerative hierarchical clustering**, in which each item is initially considered to be a **singleton cluster** of size 1. The most highly similar items are first put together forming clusters of size 2. Then the procedure looks at the resulting set of clusters again, using a slightly lower threshold of similarity and puts together clusters that are similar (above threshold) and more similar to each other than to any other clusters. If so, they are added to the cluster and another round of clustering is redone. In each round, the threshold level of similarity is slowly lowered, until a certain criterion for stopping has been reached. The criterion for stopping may be chosen according to some mathematical criterion or may simply be manually set, depending on the situation. Again, the final clustering solution may not be unique or optimal.

To improve robustness, sometimes clustering is carried out multiple times, keeping track of which items are most frequently assigned to the same cluster, and the final solution is an average or consensus of all the clustering solutions. Another way to improve robustness is to carry out **resampling**; that is, after clustering is complete, remove some of the items from a cluster, and run an extra clustering step to see if these items get reassigned to a different cluster.

There are formal criteria for evaluating clusters. For example, one can ask if a particular method maximizes the similarity of items within clusters, while minimizing similarity across clusters. However, clusters are usually evaluated by how well they provide insight. If we want to group together patients with autism according to how closely they share similar symptoms, demographics, or comorbidities (i.e., what other diseases they have, such as epilepsy), we might play around with different total numbers of clusters and different similarity metrics, to see if we can discern discrete medical subtypes of the disease. However, the clusters that emerge are exploratory and suggestive and do not comprise definitive evidence of their own.

Tip: Like we said when performing a t-test, the mere act of achieving significance is not evidence that a hypothesis is correct. Similarly, just because a machine-learning method produces a classification or clustering solution does not provide evidence that the models or the clusters have any real-world meaning. The findings are a starting point for further investigation using new data and new types of evidence.

SOME MACHINE-LEARNING METHODS

Many of the workhorse machine-learning methods, such as linear regression, logistic regression, clustering, and others discussed below, can be performed easily by using free, open source software, notably the Weka software suite http://www.cs.waikato.ac.nz/ml/index.html [3]. Weka does not require you to write computer programs but merely to understand what the parameters mean and how to choose them! This workbench also makes it easy to compare different machine-learning methods and see which appears to be best in a given situation. It is necessary to compare methods, for despite the highly mathematical nature of machine-learning procedures, there is still little theoretical understanding of how to predict which method will give the best performance on any particular problem.

Three popular machine-learning methods that we have not yet discussed are **decision trees**, **support vector machines (SVMs)**, and **neural networks**, including **deep learning**.

Decision Trees

These can be used either for supervised classification (putting examples into one or more categories) or regression (estimating numerical values or probabilities). As the name implies, the underlying algorithm processes each example through a sequence of yes/no decisions. Fig. 13.7 shows a decision tree that outlines features that predict whether a passenger on the *Titanic* will be a survivor or not. Passengers who were male, over 9.5 years of age, and had three or more family members (sibsp = spouses or siblings) were most likely to survive.

Support Vector Machines (SVMs)

These are a supervised classification method. SVM finds a line, plane, or hyperplane (depending on the number of independent plus dependent variables) that optimally *separates* positive examples from negative examples (or more generally, separates group 1 from group 2). Note that this is not finding a best-fit line but finding a best separating line!

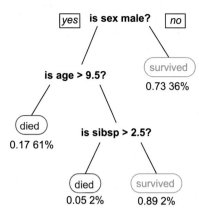

FIGURE 13.7 Decision tree for *Titanic* passenger survival. The figures under the leaves show the probability of outcome and the percentage of observations in the leaf. *Figure prepared by Stephen Milborrow, accessed from https://commons.wikimedia.org/w/index.php?curid=14143467#filehistory and used with permission.*

The problem of finding a best separating line is made more interesting when the two groups simply cannot be separated by a line or plane when plotted in their original dimensions. For example, Fig. 13.8, panel at left, shows two groups that form rings in a two-dimensional graph. Clearly, no line can separate two concentric circles! Yet it is possible to transform the datapoints into three-dimensional space in such a way that the two groups can be cleanly separated by a plane (Fig. 13.8, panel at right). Because these transformations

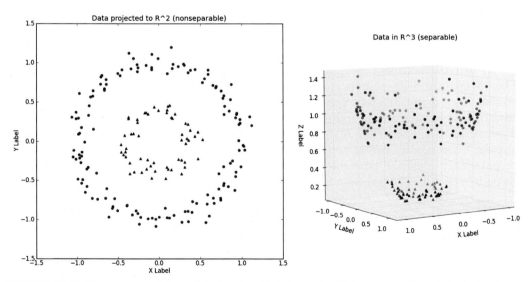

FIGURE 13.8 Two groups of datapoints (rendered in blue and red) that are not linearly separable in two dimensions (at left) but become separable when the data are transformed into a new three-dimensional representation (at right). *Figure created by Eric Kim http://www.eric-kim.net/eric-kim-net/posts/1/kernel_trick.html and reprinted with permission.*

often involve functions called **kernels**, this is referred to as the **kernel trick** and is a leading reason for high performance of SVM methods. It is usual to try several different types of kernels (e.g., so-called polynomial, Gaussian/radial basis function, or sigmoid kernels), to see which performs best for a given data set.

Neural Networks and Deep Learning

Neural networks were introduced by McCulloch and Pitts [4]. They envisioned idealized neurons, each receiving inputs on their dendrites, summing up voltages as passing a threshold or not and sending out signals to neighboring neurons. They showed that such a network could, in principle, compute any function that a Turing machine could compute—meaning that a neural network has the same power as the digital computer. In fact, historically, John von Neumann used the McCulloch–Pitts neuronal network as the basis for designing the digital computer.

Neural networks led the way in early attempts at making machines that can think intelligently (Artificial Intelligence [5]) during the 1950s and early 1960s. However, computers of that era were not adequate to support the multiple layers of neurons and the complicated algorithms necessary for them to be practical. In the 1970s and 1980s, a movement in psychology known as connectionism emphasized the cognitive importance of distributed networks of idealized neurons. Finally, in the 1990s and 2000s, computers and computer science itself had advanced to the point that very large and complex artificial neural networks could be constructed and used fruitfully. The availability of graphical processors (which carry out computations very quickly in parallel, rather than one at a time) was one key breakthrough, as was the use of truly massive amounts of training data (billions of examples). The so-called backpropagation algorithm allows one to implement reinforcement training by adjusting the strength of neuronal connections throughout the entire network. Today, the study of deep learning is one of the hottest and fastest moving areas not only in machine learning but in all of science. Deep learning methods are being used for facial recognition, translating foreign language text, even creating computer-drawn art.

Despite the (justified) excitement about neural networks, they are incredibly complex entities that have many different possible choices regarding layers and architectures, many different possible parameters, and many different training strategies that need to be tuned and optimized. They also have many hyperparameters, which are variables that need to be set by the experimenter before actually optimizing the model's parameters—for example, hyperparameters describe how the network should be trained. Much simpler to create a decision tree or a linear regression whenever possible!

BIG DATA

Big Data is not hard to find. Think of all the cell phone calls around the world, including both their metadata—who called who, when, for how long, from where (closest cell tower or GPS)—and their content (text, pitch, duration, accents, voice identities...). Think of all the tweets on Twitter, all the posts on Facebook, all the Youtube videos—indeed, think of all Internet content whatsoever. Think of all of the data contained in a photographic survey of outer space and all the information collected regarding particle collisions in an atom smasher. Think of all the temperature measurements as they are recorded every day, in every part of the world.

Most of these examples involve streaming data, rather than large static databases. This means that the data that you might want to analyze are not sitting in one place, but are constantly changing character and exhibiting trends that vary over time. Modeling graphs and networks are key techniques used in analyzing Big Data, as well as creating and factoring multidimensional arrays of numbers arranged as matrices and tensors (tensors are multidimensional matrices and fall out of the scope of this book).

One of the big potential gains to be had from Big Data is to **form new knowledge by combining different types of data together**. This can be challenging, however, since each source of data has its own properties and its own meaning. One cannot simply merge different data sets. Instead, shared reference points can be used to "sew" the data sets together. Both commercial entities, such as Facebook and Google, and national intelligence agencies combine information from multiple sources to make individualized profiles of a person's friends, likes, buying habits, and so on. However, when data are not properly "sewn" together, embarrassing mistakes can happen: a Range Rover mentioned in one source may be a vehicle, whereas a Range Rover mentioned in a different source may refer to a cowboy.

Another important technique to form new knowledge comes from mining **implicit information**. This is information that is not explicitly stated but can be inferred. For example, suppose you have a database of individuals containing their first and last names as well as, say, the record of their credit card purchases over the past year. You would like to analyze whether males and females show different buying behavior, but you have no explicit information saying who is male and who is female. Yet, you can generally infer a person's gender from first names with high accuracy. And persons predicted to be female according to first name, who have also purchased feminine hygiene products, are especially likely to be female. This type of inference allows you to create a new feature called "imputed gender," which can assign the most likely gender to each individual. Then, in turn, the imputed gender field can be used to study and factor the buying differences between males and females. A few errors will occur using this strategy, yet the analysis of buying behavior across genders may be very accurate on a large scale.

Another way to make use of implicit information is to take an item and make a list of the other items that are most closely associated with it (e.g., that cooccur most frequently in some context, or are most similar according to some metric). Then, the list can be used as an **implicit representation** of the item itself. We will see an example of this in topic modeling (see below).

DIMENSIONAL REDUCTION

There are many reasons to reduce the number of dimensions in a scientific problem. For example, it is often desirable to choose only the most important factors that predict a given outcome (i.e., those independent variables that account for the bulk of the variability in the outcome variables) and discard the rest. This focuses and simplifies the analysis and often improves the performance of predictive models substantially.

Another type of reduction occurs when one feature has many dimensions and it is transformed, or mapped, into a smaller number of new dimensions. For example, a personality profile questionnaire may have 500 questions, but they are not all independent of each other. They may be probing just five or six major underlying personality factors. These constructs are called hidden or **latent** factors. We believe that latent factors exist in some sense (we believe that personality factors such as extroversion vs. introversion have some basis in reality). However, latent factors are not directly observable on their own but rather are comprised of a combination of observable features.

There are a dozen or more dimension reduction methods commonly employed by scientists, but I will mention only two here:

Principal component analysis is a method for identifying latent "features" that explain most of the variability in outcome measure(s). The one that explains the most variability is the first principal component, the one that explains the next most is the second principal component, and so on. The method relies on the same linear model used in multiple regression

$$y = k + ax_1 + bx_2 + cx_3 + \ldots + \varepsilon$$

except that "features" are not independent variables; rather, each principal component is **a weighted sum of several independent variables**. The method is performed so that each principal component is completely independent of (uncorrelated with) each of the others.

Principal component analysis is a very popular technique, because it gives a quick-cut summary of most important influences in an experiment or data set. However, it may be difficult to interpret what the principal components actually stand for in scientific terms.

For example, suppose you are carrying out an experiment measuring a total of 40,000 mRNAs and you want to identify the most important changes that occur in one condition versus another. The most robust changes may not be the mRNAs that change the most. Rather, using principal component analysis, you might identify a small group of (say) 12 mRNAs that comprise the first principal component of the response, accounting for most of the variability in expression across the two groups. It is wonderful to focus on 12 mRNAs rather than 40,000! But looking at the 12 mRNAs, it may not be obvious how they form a coherent set—they may or may not belong to the same pathway, or even be regulated together—so the biological meaning of the first principal component may remain unclear, even if it is useful in characterizing the response.

Topic modeling: If you consider each word to be an independent variable, there are many thousands of different words found in a typical document, so representing a document in terms of its words requires a vector that has thousands of dimensions. Think of the words and documents as creating a word:document matrix, where different documents are listed on rows and different words are listed across thousands of columns. The entry on (row 2, column 3) will say how many times the word 3 appeared in document 2. This can be very cumbersome to handle. Matrix factorization can be used to break this into two smaller matrices—a word:topic matrix and a topic:document matrix, where the **topics** correspond to the **principal components** of the word:document matrix. Each topic represents a combination of words that are most closely associated with each other. For example, one topic may consist of the words (bank, money, teller, interest) and another topic may consist of the words (bank, river, trees, picnic). By keeping just the first few principal components, one can summarize the most important topics mentioned within a given document as well as across the documents.

Even though you may never yourself employ one of these machine-learning or data mining methods, you should understand the basic ideas of how they work, since they are used very widely in research reports across the physical, biological, and social sciences, and even reported in the popular press. Nate Silver's blog, Five Thirty Eight (https://fivethirtyeight.com/), is a great place to read about statistics and data mining as they apply to sports, politics, and science.

References

[1] Theobald R, Freeman S. Is it the intervention or the students? Using linear regression to control for student characteristics in undergraduate STEM education research. CBE Life Sci Educ Spring 2014;13(1):41–8. http://dx.doi.org/10.1187/cbe-13-07-0136.

[2] James G, Witten D, Hastie T, Tibshirani R. An introduction to statistical learning. New York: Springer; February 11, 2013.

[3] Frank E, Hall MA, Witten IH. The WEKA workbench. Online appendix for data mining: practical machine learning tools and techniques. 4th ed. Burlington, Massachusetts: Morgan Kaufmann; 2016.

[4] McCulloch WS, Pitts W. A logical calculus of the ideas immanent in nervous activity. Bull Math Biophys December 1, 1943;5(4):115–33.

[5] Russell R, Norvig P. Artificial intelligence: a modern approach. 3rd ed. Prentice Hall; 2010.

MAKE YOUR DATA GO FARTHER

14

How to Record and Report Your Experiments

SCIENTISTS KEEP DIARIES TOO!

We live in an age of atom smashers, moon rockets, smartphones, and self-driving cars. Yet almost everyone (in the academic world, at least) still uses pen and paper to keep their experimental notes, writing in a hard-bound notebook that has carbon paper to allow separate storage of the carbon copies (of course, the experimenter often scotch-tapes in photos, printouts, and other items, which may not be preserved as copies at all). The secrecy, intimacy, and sheer haphazard sloppiness of lab notebooks is much more akin to a diary than it is to any kind of formal or official record.

Electronic lab notebooks do exist—in fact, there may be as many as 100 different notebooks on the market. Some notebooks are expensive, require monthly subscriptions, and are oriented for large industrial laboratories that are worried about documentation for patents and intellectual property. Some notebooks are **open source** (meaning the computer code is public and can be modified by users) and free or low cost (e.g., sciNote or eLabFTW); some are Web based to foster collaboration; some can be operated from tablets and smartphones; some have links to statistical software. Most have date and **version control**, meaning each change in a lab note is documented and saved as a separate version. A specialized type of notebook, called a **workflow**, will be discussed in the next chapter. Some electronic notebooks are specific for domains such as chemistry, whereas others are generic and can be used in any field (e.g., Microsoft OneNote, Google Drive, or Wiki platforms). Yet none of these options has yet reached the tipping point in general acceptability–and I am not entirely certain why not.

Regardless of how experiments are documented, the lab notes must satisfy a number of constraints and requirements [1]:

How Long to Keep Lab Notes

First and foremost, keep in mind that lab records must be kept for at least, let us say, 7 years. There is no magic number that everyone agrees on, but the notes need to be available in case you are audited by your funding agency or university, or accused of scientific

Data Literacy
http://dx.doi.org/10.1016/B978-0-12-811306-6.00014-2

misconduct or simple error. This means keeping your notes for a certain number of years beyond the end of the grant period, or beyond the date of publication. You should also keep notes available for other members of your own laboratory, who may be trying to replicate results, check details of methods and protocols as part of their follow-up studies, or use the raw data for further analyses. In practice, I keep my notes indefinitely (over 30 years).

Moreover, you should not underestimate the risk of being accused of scientific misconduct or error of one kind or another. The only protection is to have meticulous records of the raw data including what was done, by whom, and on what dates. This means having an audit trail of both the data as it is handled and transformed, i.e., the **data provenance**, as well as the notes and associated readouts, protocols, data sheets, etc. A paper that I was a coauthor on was recently attacked by a blogger claiming that it was full of misstatements and rookie data analysis errors. Fortunately, the raw data was not only kept but had been placed in a public archive. We were able to respond to each of the claims, of course, but the effort needed to respond, carefully and fully, took months of our time [2].

If the research has any intellectual property or patent implications, the notebooks need to be kept for inspection for lawsuits that may be filed well after the original patent is submitted or granted, and there needs to be timestamps to prove when the notes were written, as well as a way for witnesses to sign the notebooks.

Since notes are often written over an extended period of time, and items may be left blank and filled in, or mistakes corrected later, an electronic notebook needs to be able to maintain multiple versions. It is not enough to keep the most recent version; the trail of changes and corrections also need to be reconstructable.

Where and How to Store Lab Notes

Of course, the original notes should be stored in secure (locked) locations, and carbon copies and backup electronic files should be stored separately, *not* in the same laboratory. I personally had my laboratory subjected to a massive flood when pipes running along the ceiling froze and then burst in mid-winter, ruining not only the notebooks but also the computers and equipment that had data stored electronically on them.

Remember that electronic storage formats change constantly (Fig. 14.1) (punch cards giving way to the reel-to-reel tape giving way to eight-track tapes to cassette tapes to compact discs to data sticks to…), and the same is true for operating systems and word processing and spreadsheet software, which undergo upgrades on a regular basis—and upgrades may not all be backward compatible with earlier versions. Thus, looking ahead 30 years, it is wise to store text files, or at least backup versions of your text files, not only in proprietary formats such as Word but in generic formats that can be read by any text editor (e.g., files that have the extension .txt, or using open source software such as OpenOffice). Similarly, instead of only saving spreadsheets in Excel, it is wise to have additional backup copies saved in more generic formats (such as the extension .csv).

Not only should you expect to encounter fire, flood, acid spills, and the like during your careers, but also there is an even greater risk of encountering disgruntled, unbalanced, unethical, spiteful, or frankly psychotic students and coworkers who may deliberately attempt to alter or destroy your notes. I have seen each of these types of colleagues come through my

FIGURE 14.1 **How many of you reading this were alive in the 1970s, when tapes were state of the art?**

department over the course of my career. James Holmes, who carried out a mass shooting in Colorado in 2012, was a neuroscience graduate student.

One of the awkward aspects of keeping lab records is the fact that not all items are stored in the same place or in the same format. When a vial of antibody is ordered from a commercial source, the actual antibody is stored in a refrigerator or freezer; since these are perishable, the method and time of storage and even details such as whether the antibody was diluted or aliquoted are relevant to note but are not usually written anywhere. The data sheet describing the antibody is usually present as a hard copy, including both the catalog number and lot number, whereas some other details may only be available on the company website. The protocol describing how the antibody is used in a given experiment and any supporting information validating the use of the antibody for purposes, such as immunoblotting or immunoprecipitation, are placed by the investigator in the lab notes, possibly in specified locations by category, possibly written chronologically so that they are hard to search for or find again. Finally, the actual experiments that employ antibody are described in the notes. Linking information about the specific lot of antibody across these different locations is hard enough, but consider that this situation applies to every reagent, lab resource, software, or type of data that is acquired (which may include photographs, X-rays, fMRI or other electronic images, numerical printouts, sequence files, and so on). Cross-referencing all these with the notes over time, and keeping track of changes such as which version of software was used for which specific experiments, is a major headache and hassle.

WHO OWNS YOUR DATA?

I do not know who owns your data—do you? If you are a student working on a school project or a PhD thesis, you probably own anything that you generate. Larry Page and Sergei Brin were graduate students at Stanford when they developed the initial ideas and software that led to Google; they, not Stanford or their thesis advisors, owned the intellectual property. However, if you are a student working on a grant, or hired to work on someone's project, chances are that they, not you, own the ideas and data that you generate. It is worth knowing.

If you are a faculty member, you may or may not own anything that you generate both as an employee and on your own time. Academic freedom is the idea that faculty can say, write, or teach freely, without censorship or the need to obtain approval from anyone (although they take personal responsibility for upholding academic standards). However, if you are a faculty member working on a grant or contract, chances are that the grant or contract was given to the university and they, not you, own the ideas and data that you generate. Depending on the terms of the grant or contract, the funding agency may or may not have the right to preapprove publication of your findings. This is a very slippery and confusing topic and varies from one institution to another. Moreover, as academic freedom and tenure have become under threat, and as universities seek more actively to commercialize the inventions made by their faculty, the ownership of faculty output is becoming less clear over time.

When a scientist submits a manuscript for publication, traditionally the journal asks for transfer of **copyright** from the scientist to the journal. This allows the journal to publish the article, charge for reprints, restrict access to subscribers, etc., generally without paying the author anything. The author still retains ownership of the ideas and findings described in the article, but may lose the right to post the article publicly; and people reading their article may not have the right to download or circulate it to others, to use the figures or data, or even to reformat the article. This has led to a backlash against traditional publishers and fostered the open access movement, which is described in Chapter 16.

REPORTING AUTHORSHIP

The average number of authors per paper, across all fields, is around five and is steadily rising over time. Even in mathematics, the majority of published scientific articles have two or more authors. In biomedical sciences, laboratories collaborate widely, so that coauthors may come from different institutions or even different countries. Generally, scientists work together without any formal written contracts or agreements about authorship. This works well most of the time, although it is worth approaching this issue carefully, since disputed authorship is one of the common reasons for retracting a published article.

One way that a dispute can arise when a paper is published without one or more of the authors being aware is that their name is listed as coauthor. The accepted standard is that all people listed as coauthors should attest that they have seen and approved of the final version being submitted and give written permission in advance that they agree to be coauthors. Sounds reasonable, but often labs are in a hurry to publish and the submitting author skips this step. This has actually happened to me more than once in my career—I assisted in work that was primarily done in some other laboratory, but was not notified in advance that they were adding my name as coauthor, and I was not given the opportunity to read the final article before it was submitted. I believe that the lab chiefs involved in these cases felt that they were doing something to make me happy. They could have mentioned my name in the acknowledgments section without giving me authorship—that would not require me to read their manuscript or approve it. And I did fulfill the criteria for being an author. I did not protest being added to their papers, but you, the reader, should know that this is not good scholarly practice. And there are cases where a researcher submitted an article to one journal, without realizing that his or her lab chief had submitted the same work to a different journal!

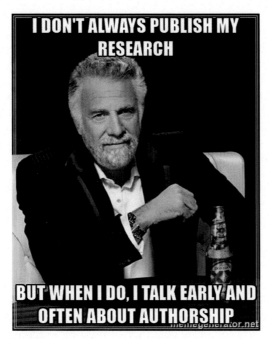

FIGURE 14.2 **More good advice from the Most Interesting Man in the World.** *Reprinted from https://memegenerator.net/instance/31651105.*

Tip: Always send the final version of a manuscript to all coauthors and get written permission from them before submitting. It is also wise to inform people who you plan to include in the acknowledgments (but not include as coauthor) to avoid confusion and misunderstandings (Fig. 14.2).

The person submitting an article is called the **corresponding author**, and in most cases, is the person who takes responsibility for the entire work; he or she is the one who should be contacted for questions, data, or any other issues related to the article (such as ethical issues). Sometimes more than one person is designated as corresponding author. Rarely, someone else may submit on the corresponding author's behalf—this may happen if a scientist dies before the article has been submitted, or if there is a submission deadline that happens while the corresponding author is traveling abroad. The head of the laboratory where the research was performed is often corresponding author, but that is not a fixed rule, and if a student or staff scientist carried out the bulk of the research, they may be designated as corresponding author as long as they are in a position to take responsibility for the overall study.

The corresponding author usually is the one who decides who else to list as coauthor. The accepted standard is that **anyone who provides significant intellectual contribution to the study should be offered the opportunity to be a coauthor**. Someone who did a lot of routine work (e.g., a technician who prepared and assayed samples, or a data entry clerk who crunched the numbers) is not eligible unless they made some intellectual contribution (e.g., modified a protocol in a critical way). Someone who solely provided financial support for the project (e.g., the department head or other senior mentor) is not eligible. Someone from another laboratory who lent reagents which were critical to the success of the study may or may not qualify for authorship: If they simply gave the reagent, they should be acknowledged in the acknowledgments and their reagents should be cited. However, if they provided guidance on how to use the reagents, or gave other unpublished information or expertise, then they arguably deserve to be offered coauthorship. Again, I say "offered" coauthorship, because if they do not agree with how the study was done, and its conclusions, or if they simply feel that their contribution was not great enough, they may well decline to be coauthors.

Actually, only the corresponding author must stand behind every aspect of the published article. The other authors only need to vouch for the part of the study that they contributed. Many, perhaps most, journals now include a section in which each coauthor's role(s) on the paper are specified. A new suggested taxonomy of roles called CRediT [3] makes it easy to specify which author(s) were involved in conceptualizing the study, provided methods or software, performed the experiments, wrote the initial draft, and so on. The listed roles even include activities that do not, by themselves, justify authorship, such as obtaining funding for the work.

Listing roles explicitly protects the coauthors in several ways:

1. It shields them from responsibility for mistakes done by others. Historically, each of the authors on an article has individually been considered responsible for vouching for its accuracy. When articles have been found to involve fraud or scientific misconduct, this has tainted the reputation of all coauthors. Fraud by students has even led to the resignation of their advisors, who ought to have provided better oversight, even if they were not the corresponding authors. However, as research more often involves large collaborative studies across multiple laboratories, it has become less practical to expect that each participant can check and cross-check the work of everyone else.
2. Having a checklist of roles encourages the corresponding author to include people as coauthors who otherwise might be overlooked (e.g., those who write code for the project are giving intellectual input!).
3. It helps with the overall assignment of credit for each coauthor. When a paper is published with nine coauthors, how should credit be distributed? Should each coauthor get credit for writing one paper? (After all, each of them puts the paper on their CVs.) Should each coauthor get one-ninth of a paper? Does it even make sense for credit to be fractional and add up to one? Just as scientific articles vary in their importance and impact, so the role of each coauthor needs to be considered on their individual merits.
4. Specifying roles is much better than observing the order in which the authors are listed on the paper. In mathematics, authors are usually listed in alphabetical order. However, norms vary across fields, and there is no general rule. In biomedical sciences, the first author (also called the **senior author**) is often a student or postdoctoral fellow who did the bulk of the work and wrote the initial draft, whereas the final author is often the corresponding author, who may be the student's advisor or lab chief.
5. Finally, this should diminish the pernicious practice known as **ghost authorship**. This occurs when the people actually doing the study, and/or the people writing up the study, are NOT listed as authors on the article at all. Instead, famous or prominent experts are induced (or paid) to allow their names to be put on the article as authors,

thus giving a boost in visibility, credibility, and apparent objectivity. Ghost authorship is most problematic in the pharmaceutical industry, involving articles that report clinical trials of drugs. Sometimes one clinical trial will write up multiple papers that do not share any authors among them, thus giving a false impression that they come from independent sources of evidence. Another type of ghost authorship involves review articles that summarize evidence about drug studies, in which the article is not written by the expert whose name is on it. The expert may receive a high fee for reading the manuscript and allowing their name to be placed on the paper.

REPORTING CITATIONS

In some ideal sense, an author cites previously published work to declare what studies influenced his or her thinking, spurred the rationale, or led to or altered the course of the study that is being reported. That is, the citations might be thought to provide an accurate, honest record of the intellectual influences on the present paper.

In practice, however, citations are not created in that way at all. What usually happens is that the scientist completes the study, and during the writing phase, does **due diligence**, which is a search to find related research (including work that they were not previously aware of), and makes sure to cite all relevant prior studies—especially any that were written by people who are likely to serve as reviewers for the journal! This approach is more scholarly, gives better context to the reader, and gives credit for the work of others. These are all good things, as long as you recognize that the list of citations does not accurately report which work was influential in the scientist's thinking at the time the work was done.

Journals vary in what types of documents they cite and how they cite. Published journal articles, online journal articles, books, and conference proceedings are generally included in a list at the end of the article, either in alphabetical order or numbered, and these comprise the formal References of the article. Sometimes journals will cite links to websites; links to articles posted online (not in a peer-reviewed journal); or personal communications (these generally require written permission from the person being quoted).

Many of the valuable contributions to a study remain uncited in the traditional scheme and are often not even clearly described within the text of the article itself. For example, commercial reagents such as antibodies may be critical to the outcome of a study, but previous studies that characterized or used the same antibody are usually not cited. And describing the antibodies in a paper by vendor and even catalog number is not enough; one needs to specify the lot number and ideally one should identify its entry within an external database such as The Antibody Registry (http://antibodyregistry.org/) [4]. Besides reagents, there are emerging efforts to formalize how to describe and cite data sets and software (including the version and the specific parameters used). Citing these resources does a lot more than simply help the originators get their share of credit. It also helps scientists find other articles that used the same resources.

A common practice in the literature is to cite websites by using their links or URLs (i.e., addresses that start with http:// or https://). However, website content may change constantly [5]. When citing a URL, one should always give the date when a website or database was accessed. Links themselves are very unstable; the phenomenon is known by the charming name of "link rot." Studies have shown that link rot occurs at a dizzyingly high rate, so that within 5–10 years, half the links cited in published papers may have been lost [6–8]. Journal articles and data that have been assigned permanent Digital Object Identifiers, or DOIs, are in much better shape since DOIs are truly permanent and take versioning into account. DOIs are backed up in multiple locations and are independent of the physical location of the data, which is not generally true of most Web content indexed by URLs.

Archiving the Web page is another approach. A nonprofit organization called Internet Archive saves a wide range of websites, but by no means most of them; and the version that they archive may not be the version that is cited by any given author. An initiative called WebCite archives specific Web pages on request from individual authors, but this does not capture the dynamic back-end functions of servers that maintain query interfaces and run algorithms on their back ends. Bear in mind that a published article that displays a URL may not indicate if that website has been archived or if it has a permanent link.

WRITING THE INTRODUCTION/MOTIVATION SECTION

We now turn to tips for writing the different narrative portions of a scientific article. The Introduction section introduces a scientific question or problem; gives a brief overview of its history, scope, and background; explains what the current state of knowledge is; and explains how the present paper will tackle or contribute to a gap in that knowledge. The authors may pose a testable hypothesis in the Strong Inference sense—pose a well-defined question that can be answered yes or no—or (most often) simply gather descriptive data that provides evidence, which can be interpreted as more or less consistent with a particular model.

While the Introduction is reasonably straightforward to write, scientists realize that often, the best way to frame the research project is grasped only in retrospect after the study has been completed and the writing starts! Should authors be candid about false starts, hunches, and other warts-and-all aspects that contributed to their study? Or should they "sanitize" a paper so that it sounds logical, even if it is not a historically or intellectually accurate report of how and why the experiment was done? Journal editors and reviewers generally prefer a tidy story—and there IS such a thing as Too Much Information!—yet I believe that honesty is usually the best policy and actually makes a more interesting story.

Another question concerns **self-plagiarism**, which means reusing entire sentences or paragraphs from your own earlier articles. Most people agree that it is OK to reuse text passages that describe methods or protocols, but NOT OK to reuse paragraphs in the Introduction, Results, or Discussion. Even when an article is part of a series, it should read as if the author is taking a fresh look at the problem.

WRITING THE METHODS SECTION

When journals were exclusively published in print form, and space was limited, it was virtually impossible to write down the methods in adequate detail so that others could replicate all of the crucial details. Now, even print-based journals generally have an online presence, and it is possible (actually, it is imperative) to write out the complete methods, even if these are placed in an online supplement. I have seen papers in *PNAS* whose online supplements are five times as long as the printed article.

The scientific community has attempted to help in research reproducibility by creating **standardized reporting guidelines** for common types of studies:

Animals

The **ARRIVE guidelines** were first proposed in 2010 [9] and have been widely adopted by biomedical journals and funding agencies (Box 14.1). Although the guidelines may seem self-

---

BOX 14.1

THE ARRIVE GUIDELINES [9]

Title

Provide as accurate and concise a description of the content of the article as possible.

Abstract

Provide an accurate summary of the background, research objectives, including details of the species or strain of animal used, key methods, principal findings, and conclusions of the study.

Introduction

Background

1. Include sufficient scientific background (including relevant references to previous work) to understand the motivation and context for the study and explain the experimental approach and rationale.
2. Explain how and why the animal species and model being used can address the scientific objectives and, where appropriate, the study's relevance to human biology.

Objectives

Clearly describe the primary and any secondary objectives of the study, or specific hypotheses being tested.

Methods

Ethical Statement

Indicate the nature of the ethical review permissions, relevant licenses (e.g., Animal [Scientific Procedures] Act 1986), and national or institutional guidelines for the care and use of animals, that cover the research.

Study Design

For each experiment, give brief details of the study design including the following:

1. the number of experimental and control groups;
2. any steps taken to minimize the effects of subjective bias when allocating animals to treatment (e.g., randomization procedure) and when assessing results (e.g., if done, describe who was blinded and when);
3. the experimental unit (e.g., a single animal, group or cage of animals); a timeline diagram or flow chart can be useful to illustrate how complex study designs were carried out.

Experimental Procedures

For each experiment and experimental group, including controls, provide precise

Continued

BOX 14.1   *(cont'd)*

details of all procedures carried out. For example:

1. how [e.g., drug formulation and dose, site and route of administration, anesthesia and analgesia used (including monitoring), surgical procedure, method of euthanasia]. Provide details of any specialist equipment used, including supplier(s);
2. when (e.g., time of day);
3. where (e.g., home cage, laboratory, water maze);
4. why (e.g., rationale for choice of specific anesthetic, route of administration, drug dose used).

Experimental Animals

1. Provide details of the animals used, including species, strain, sex, developmental stage (e.g., mean or median age plus age range), and weight (e.g., mean or median weight plus weight range).
2. Provide further relevant information such as the source of animals, international strain nomenclature, genetic modification status (e.g., knock-out or transgenic), genotype, health/immune status, drug or test naïve, previous procedures, etc.

Housing and Husbandry

Provide details of the following:

1. housing (type of facility, e.g., specific pathogen free (SPF); type of cage or housing; bedding material; number of cage companions; tank shape and material, etc., for fish);
2. husbandry conditions (e.g., breeding program, light/dark cycle, temperature, quality of water, etc., for fish, type of food, access to food and water, environmental enrichment);
3. welfare-related assessments and interventions that were carried out prior to, during, or after the experiment.

Sample Size

1. Specify the total number of animals used in each experiment, and the number of animals in each experimental group.
2. Explain how the number of animals was arrived at. Provide details of any sample size calculation used.
3. Indicate the number of independent replications of each experiment, if relevant.

Allocating Animals to Experimental Groups

1. Give full details of how animals were allocated to experimental groups, including randomization or matching if done.
2. Describe the order in which the animals in the different experimental groups were treated and assessed.

Experimental Outcomes

Clearly define the primary and secondary experimental outcomes assessed (e.g., cell death, molecular markers, behavioral changes).

Statistical Methods

1. Provide details of the statistical methods used for each analysis.
2. Specify the unit of analysis for each data set (e.g., single animal, group of animals, single neuron).
3. Describe any methods used to assess whether the data met the assumptions of the statistical approach.

BOX 14.1 *(cont'd)*

Results

Baseline Data

For each experimental group, report relevant characteristics and health status of animals (e.g., weight, microbiological status, and drug or test naïve) prior to treatment or testing (this information can often be tabulated).

Numbers Analyzed

1. Report the number of animals in each group included in each analysis and report absolute numbers (e.g., 10/20 and not 50%).
2. If any animals or data were not included in the analysis, explain why.

Outcomes and Estimation

Report the results for each analysis carried out, with a measure of precision (e.g., standard error or confidence interval).

Adverse Events

1. Give details of all important adverse events in each experimental group.
2. Describe any modifications to the experimental protocols made to reduce adverse events.

Discussion

Interpretation/Scientific Implications

1. Interpret the results, taking into account the study objectives and hypotheses, current theory, and other relevant studies in the literature.
2. Comment on the study limitations including any potential sources of bias, any limitations of the animal model, and the imprecision associated with the results.
3. Describe any implications of your experimental methods or findings for the replacement, refinement, or reduction (the 3Rs) of the use of animals in research.

Generalizability/Translation

Comment on whether, and how, the findings of this study are likely to translate to other species or systems, including any relevance to human biology.

Funding

List all funding sources (including grant number) and the role of the funder(s) in the study.

---

evident (almost to the point of "Duh!"), in fact, these recommendations are a big, big step forward, because most animal studies had not included more than a fraction of the relevant items listed in the Box. Alas, scientists do not seem to be adhering to the guidelines yet [10], which may reflect reluctance both to change the way they design their experiments as well as to change the way they write up their articles.

Animal experimenters are especially deficient in reporting three aspects:

1. The method of randomization is generally not stated, although this is a basic, crucial element of design. Since animals belonging to the same litter are not independent of each other, it is specifically important to specify how randomization took litters into account.

2. Blinding is generally not stated either. However, ideally the experimenter handling the animals should be blind to which treatment the animal is getting; the person rating the animal response should definitely be blinded; and the person analyzing the data should be blinded until the data analysis is complete. Unless these are specifically stated in the paper, it is likely that these safeguards were not done.
3. Censoring bias: If surgery was involved, how many animals died or were discarded (and for what reason)? If there were microinjections, how many animals had bad placements (and how was that ascertained?) or had side effects such as bleeding? These are essential parts of understanding the experiment (and assessing how much variability is due to the experimental set up) but are very rarely mentioned.

Clinical Trials

Clinical trials should follow the **CONSORT guidelines** [11], which describe recruitment, participants, randomization schemes, blinding, outcomes, dropout of subjects, adverse events, and so on. Because clinical trials tend to be larger, more formal, more long-lasting, and involve more investigators than typical animal studies, one might expect that clinical trial articles would adhere more closely to guidelines—but no, although some aspects of clinical trials are better reported post-CONSORT, many critical details of randomization, blinding, and statistical analysis are still missing in the majority of published articles, even in the "best" journals in the field [12].

Specialized research domains also have recommended guidelines for the conduct and reporting of experiments, including those for genetic association studies (GWAS), microarray studies, nucleotide sequencing, functional MRI (fMRI) data, and even studies of spinal cord regeneration in humans [13]. Looking on the bright side, even if investigators are not adequately following the guidelines, at least not yet, the proliferation of community-driven efforts means that more and more scientists are getting serious about the need to write reports that include full details.

Reporting Reagents, Protocols, Software, and Equipment

Commercial reagents are moving targets and change over time, unpredictably and with poor documentation. I am ambivalent about their use in experiments. On the one hand, it is beneficial for different laboratories to share the same reagent, which provides one level of standardization, and commercial reagents certainly make it much easier for scientists to carry out research. On the other hand, proprietary reagents are trade secrets, meaning that they do not tell you what is in it. They may not be optimal for your (or even anyone's) experiment, and it is hard for you to tinker with them. Nor do they necessarily inform customers when they change the formula—you can order the same catalog number and receive a different formulation. Reporting exactly what version of the reagent you used in an experiment becomes hard to describe.

And often the company selling the product will merge with another: The popular lipid-based transfection reagent Lipofectamine was first sold by Invitrogen, which merged with Applied Biosystems, forming Life Technologies, which was acquired by Thermo Fisher Scientific. The catalog number might not remain the same throughout these mergers, and documentation about the original reagent may be hard to trace. This is why it is important to document lot numbers, not just catalog numbers. Software and equipment generally have better documentation, and different releases have version numbers, which are important to report in an article.

Protocols are step-by-step recipes that should be specified in detail in an article. However, when the same general methods were used in prior publications in the same laboratory, there is an almost irresistible temptation to say "we performed Western blotting as done previously in reference 2." I have certainly been guilty of this, myself [14]. Why is this bad behavior? First, when you go to reference 2, you find that the protocol is not described there, either! Instead, it sends you to reference 6 and so on, in a long string that looks like a scavenger hunt. Second, most of the time, the present article does not follow reference 2 exactly, but has added some critical modification(s) which should be (but are not) described in detail.

WRITING THE RESULTS

If you have ingested all of the nuggets of wisdom I have imparted in Sections A (Experimental Design), B (Data Cleansing), and C (Statistics), then you will have no problem reporting how you cleansed and analyzed your data. In case you would like a mental nudge, the classic article, "False-positive psychology: Undisclosed flexibility in data collection and analysis allows presenting anything as significant" [15], gives explicit guidance (Box 14.2). Another useful guideline [16] is shown in Box 14.3.

I won't attempt to create another guideline (!), but I do want to highlight some crucial aspects which are missing from most published articles. For example:

1. Present a prospective power estimation, giving all parameters used in the calculations.

BOX 14.2

REPORTING OF DATA ANALYSIS

We propose the following six requirements for authors.

1. Authors must decide the rule for terminating data collection before data collection begins and report this rule in the article. Following this requirement may mean reporting the outcome of power calculations or disclosing arbitrary rules, such as "we decided to collect 100 observations" or "we decided to collect as many observations as we could before the end of the semester." The rule itself is secondary, but it must be determined ex ante and be reported.

2. Authors must collect at least 20 observations per cell or else provide a compelling cost-of-data-collection justification. This requirement offers extra protection for the first requirement. Samples smaller than 20 per cell are simply not powerful enough to detect most effects, and so there is usually no good reason to decide in advance to collect such a small number of observations.

Continued

D. MAKE YOUR DATA GO FARTHER

BOX 14.2   *(cont'd)*

Smaller samples, it follows, are much more likely to reflect interim data analysis and a flexible termination rule. In addition, larger minimum sample sizes can lessen the impact of violating Requirement 1.

3. **Authors must list all variables collected in a study**. This requirement prevents researchers from reporting only a convenient subset of the many measures that were collected, allowing readers and reviewers to easily identify possible researcher degrees of freedom. Because authors are required to just list those variables rather than describe them in detail, this requirement increases the length of an article by only a few words per otherwise shrouded variable. We encourage authors to begin the list with "only," to assure readers that the list is exhaustive (e.g., "participants reported only their age and gender").

4. **Authors must report all experimental conditions, including failed manipulations**. This requirement prevents authors from selectively choosing only to report the condition comparisons that yield results that are consistent with their hypothesis. As with the previous requirement, we encourage authors to include the word "only" (e.g., "participants were randomly assigned to one of only three conditions").

5. **If observations are eliminated, authors must also report what the statistical results are if those observations are included**. This requirement makes transparent the extent to which a finding is reliant on the exclusion of observations, puts appropriate pressure on authors to justify the elimination of data, and encourages reviewers to explicitly consider whether such exclusions are warranted. Correctly interpreting a finding may require some data exclusions; this requirement is merely designed to draw attention to those results that hinge on ex post decisions about which data to exclude.

6. **If an analysis includes a covariate, authors must report the statistical results of the analysis without the covariate**. Covariates are features which apparently have value in predicting outcomes in the study; however, they may either have direct effects on the outcome; may be correlated with the outcome but have no influence on it; or the apparent relationship may be just false positive. Reporting covariate-free results makes transparent the extent to which a finding is reliant on the presence of a covariate, puts appropriate pressure on authors to justify the use of the covariate, and encourages reviewers to consider whether including it is warranted. Some findings may be persuasive even if covariates are required for their detection, but one should place greater scrutiny on results that do hinge on covariates despite random assignment.

Modified from Simmons JP, Nelson LD, Simonsohn U. False-positive psychology: undisclosed flexibility in data collection and analysis allows presenting anything as significant. Psychol Sci November 2011;22(11):1359–66 with permission.

BOX 14.3

DATA HANDLING GUIDELINES

A Core Set of Reporting Standards for Rigorous Study Design

Randomization

- Animals should be assigned randomly to the various experimental groups and the method of randomization reported.
- Data should be collected and processed randomly or appropriately blocked.

Blinding

- Allocation concealment: the investigator should be unaware of the group to which the next animal taken from a cage will be allocated.
- Blinded conduct of the experiment: animal caretakers and investigators conducting the experiments should be blinded to the allocation sequence.
- Blinded assessment of outcome: investigators assessing, measuring, or quantifying experimental outcomes should be blinded to the intervention.

Sample-Size Estimation

- An appropriate sample size should be computed when the study is being designed and the statistical method of computation reported.
- Statistical methods that take into account multiple evaluations of the data should be used when an interim evaluation is carried out.

Data Handling

- Rules for stopping data collection should be defined in advance.
- Criteria for inclusion and exclusion of data should be established prospectively.

- How outliers will be defined and handled should be decided when the experiment is being designed, and any data removed before analysis should be reported.
- The primary end point should be prospectively selected. If multiple end points are to be assessed, then appropriate statistical corrections should be applied.
- Investigators should report on data missing because of attrition or exclusion.
- Pseudo replicate issues need to be considered during study design and analysis. [Pseudo replicates refer to measurements that are not independent of each other; they might be repeated measurements of the same sample, or measurements of samples that are related in some way and not independent of each other (e.g., come from members of the same family).]
- Investigators should report how often a particular experiment was performed and whether results were substantiated by repetition under a range of conditions.

Modified from Landis SC, Amara SG, Asadullah K, Austin CP, Blumenstein R, Bradley EW, Crystal RG, Darnell RB, Ferrante RJ, Fillit H, Finkelstein R, Fisher M, Gendelman HE, Golub RM, Goudreau JL, Gross RA, Gubitz AK, Hesterlee SE, Howells DW, Huguenard J, Kelner K, Koroshetz W, Krainc D, Lazic SE, Levine MS, Macleod MR, McCall JM, Moxley 3rd RT, Narasimhan K, Noble LJ, Perrin S, Porter JD, Steward O, Unger E, Utz U, Silberberg SD. A call for transparent reporting to optimize the predictive value of preclinical research. Nature October 11, 2012;490(7419):187−91. http://dx.doi.org/10.1038/nature11556 with permission.

2. Make it clear at the outset whether your experiment involves Strong Inference, or at least has a hypothesis with predetermined outcomes to test. Or is it an exploratory study to look for interesting patterns, with multiple comparisons done post hoc (that is, after seeing the data)?
3. Present the entire pipeline of data cleansing, including how thresholds were chosen; weird points or groups; outliers; binning; transformations; how normalizers were assessed and chosen; whether the data distributions were quasinormal, etc.
4. Include ALL the results from your study, including (1) baseline values and summary statistics for each group, (2) differences between groups, (3) ratios between groups, and (4) estimates of statistical significance of any differences that are observed.
5. Statistical tests should generally use *both parametric and nonparametric methods* (or at least defend the choice of whatever tests are used). The *P-values should be corrected for multiple comparisons* and the method of correction should also be defended. If a finding is statistically significant, that does not constitute evidence (in and of itself) that the finding is "real," important or is not subject to bias or error. Therefore, additional, independent evidence should be produced to assess and confirm any results that are being reported in the article.
6. Finally, the data are also a part of the Results and *should be attached to the manuscript or placed in a public permanent repository* along with listing an accession number or DOI. *Both raw data and cleansed data should be saved.*

Reporting Data in Figures

Data visualization (like make-up and fashion) is an art form in which it is as important to know what to conceal as what to reveal. For those of us who think in terms of text and numbers, rather than drawings and diagrams, making figures does not come naturally. However, most readers grasp graphs, histograms, bars, scatter plots, and maps much more readily than tables, and so making accurate, yet simple figures is an important part of writing a scientific article. Some guidelines for making good illustrations can be found in Ref. [17]. Here, I will focus on certain key points that are often missing in published articles:

1. When plotting datapoints on a graph or histogram, make sure the X- and Y-axes are well marked. Are the scales linear, or are you plotting on a log scale? If possible, the values at the origin should start at 0. A change from 100 to 101 will look small if you plot the values starting at 0 and scale the Y-axis between 0 and 200, whereas it may look huge if you start at 100 and scale the Y-axis between 100 and 101. Thus, consider carefully whether the visual point you are making is congruent with the magnitude of the effects being reported.
2. Error bars should almost always be attached to datapoints that summarize a group.
3. Consider carefully how you want to present the data from each group (Fig. 14.3).
   Tip: Always be clear whether the error bars indicate standard deviation (SD), standard error of the mean (SEM), a 95% confidence interval, or quartiles (when making a box plot). Similarly, it should be clear whether the datapoint is the summary data refer to mean or the median value.
4. Only show the point you want to make! Sometimes it is better to make several simple plots that each conveys one message, rather than one complicated figure. Every visual

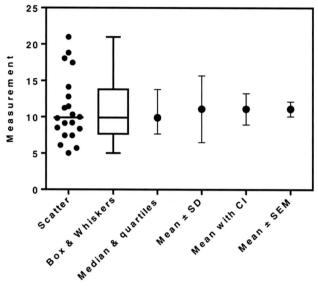

FIGURE 14.3 **Standard error bars do not show variability and do a poor job of showing precision.** Shown is one data set plotted six ways. The leftmost lane shows a scatter plot of every value, so is the most informative. The next lane shows a box-and-whisker plot (Chapter 3) showing the range of the data, the quartiles, and the median (whiskers can be plotted in various ways, and do not always show the range). The third lane plots the median and quartiles. This shows less detail but still demonstrates that the distribution is a bit asymmetrical. The fourth lane plots mean with error bars showing plus or minus one standard deviation. Note that these error bars are, by definition, symmetrical so give you no hint about the asymmetry of the data. The next two lanes are different than the others as they do not show scatter. Instead, they show how precisely we know the population mean, accounting for scatter and sample size. The fifth lane shows the mean with error bars showing the 95% confidence interval of the mean. The sixth (rightmost) lane plots the mean plus or minus one standard error of the mean, which does not show variation and does a poor job of showing precision. *Reprinted from Head ML, Holman L, Lanfear R, Kahn AT, Jennions MD. The extent and consequences of p-hacking in science. PLoS Biol March 13, 2015;13(3):e1002106. http://dx.doi.org/10.1371/journal. pbio.1002106 with permission.*

aspect of a figure needs to be considered. If something is a particular color, what does that convey? If points are placed in a particular place, what are you trying to say? Slices in a pie chart may be ordered from biggest to smallest or might be ordered differently depending on the point that is being made. Authors sometimes insert a regression line (or best-fit curve) into a scatter plot without any apparent reason, and sometimes it makes painfully clear the fact that the data are not linear or have no meaningful relationship at all.

5. Figures shown in the main body of an article can sacrifice details and precision to be readable, as long as files are included as supplementary information to validate the conclusions and to show more detailed information (in either figure or table formats).

6. When data are dense, tables can be used, but headings should be comprehensible with a minimum of explanatory legend.

7. Step-by-step workflow activities such as research protocols, clinical trials, and sequences of data cleansing steps can often be summarized using flow charts.

D. MAKE YOUR DATA GO FARTHER

FIGURE 14.4   The evolution of scientific writing.

WRITING THE DISCUSSION/CONCLUSION SECTIONS

Following the Results, the Discussion section first summarizes the main findings and places them in the context of the study. Everyone has done this. Most authors do spell out limitations of the study (such as that the results only apply to a particular population, time slice, or method of analysis). Most authors suggest implications of the work, next steps, and additional evidence or further research that is needed.

However, one key step is lacking in many articles: It is a very classy, professional, and persuasive tactic to list possible confounds and sources of bias or error and then discuss how the controls used in your study minimized or removed their influence. Furthermore, it is very powerful, rhetorically speaking, to raise alternative hypotheses or interpretations and assess whether any of these could explain your findings nearly as well as your preferred interpretation.

Most authors do not do this—perhaps they feel that the readers would not notice alternatives if they are not mentioned, or perhaps they did not think the results through in detail because the data appeared to favor their preferred interpretation. In any case, it is very persuasive to be self-critical, and accept one interpretation only after considering and discarding all others that you can think of as much less likely. Charles Darwin, in *On the Origin of Species*, explicitly raised as many objections to the theory of evolution as he could think of, and discussed each in turn [19]. He set a high bar of scholarship and integrity that is a model for the rest of us to follow (Fig. 14.4).

References

[1] https://oir.nih.gov/sites/default/files/uploads/sourcebook/documents/ethical_conduct/guidelines-scientific_recordkeeping.pdf.
[2] See Reader Comments in: De Groote SL, Shultz M, Smalheiser NR. Examining the impact of the National Institutes of Health public access policy on the citation rates of journal articles, PLoS One 2015;10(10):e0139951.
[3] Brand A, Allen L, Altman M, Hlava M, Scott J. Beyond authorship: attribution, contribution, collaboration, and credit. Learned Publishing 2015;28(2):151—5.
[4] Bandrowski A, Astakhov V, Grethe J, Martone M. An antibody registry for biological sciences. In: Front. Neuroinform. Conference abstract: 4th INCF congress of neuroinformatics, vol. 67); 2011. http://dx.doi.org/10.3389/conf.fninf.

[5] Jones SM, Van de Sompel H, Shankar H, Klein M, Tobin R, Grover C. Scholarly context Adrift: three out of four URI references lead to changed content. PLoS One 2016;11(12):e0167475. http://dx.doi.org/10.1371/journal.pone.0167475.

[6] Wren JD. 404 not found: the stability and persistence of URLs published in MEDLINE. Bioinformatics March 22, 2004;20(5):668–72.

[7] Habibzadeh P. Decay of references to web sites in articles published in general medical journals: mainstream vs small journals. Appl Clin Inform October 2, 2013;4(4):455–64. http://dx.doi.org/10.4338/ACI-2013-07-RA-0055.

[8] Hennessey J, Ge S. A cross disciplinary study of link decay and the effectiveness of mitigation techniques. BMC Bioinformatics 2013;14(Suppl. 14):S5. http://dx.doi.org/10.1186/1471-2105-14-S14-S5.

[9] Kilkenny C, Browne WJ, Cuthill IC, Emerson M, Altman DG. Improving bioscience research reporting: the ARRIVE guidelines for reporting animal research. PLoS Biol 2010;8(6):e1000412. http://dx.doi.org/10.1371/journal.pbio.1000412.

[10] Avey MT, Moher D, Sullivan KJ, Fergusson D, Griffin G, Grimshaw JM, Hutton B, Lalu MM, Macleod M, Marshall J, Mei SH, Rudnicki M, Stewart DJ, Turgeon AF, McIntyre L, Canadian Critical Care Translational Biology Group. The devil is in the details: incomplete reporting in preclinical animal research. PLoS One November 17, 2016;11(11):e0166733. http://dx.doi.org/10.1371/journal.pone.0166733.

[11] Altman DG, Moher D, Schulz KF. Improving the reporting of randomised trials: the CONSORT statement and beyond. Stat Med November 10, 2012;31(25):2985–97. http://dx.doi.org/10.1002/sim.5402.

[12] Kuriyama A, Takahashi N, Nakayama T. Reporting of critical care trial abstracts: a comparison before and after the announcement of CONSORT guideline for abstracts. Trials January 21, 2017;18(1):32. http://dx.doi.org/10.1186/s13063-017-1786-x.

[13] Anderson DK, Beattie M, Blesch A, Bresnahan J, Bunge M, Dietrich D, Dietz V, Dobkin B, Fawcett J, Fehlings M, Fischer I, Grossman R, Guest J, Hagg T, Hall ED, Houle J, Kleitman N, McDonald J, Murray M, Privat A, Reier P, Steeves J, Steward O, Tetzlaff W, Tuszynski MH, Waxman SG, Whittemore S, Wolpaw J, Young W, Zheng B. Recommended guidelines for studies of human subjects with spinal cord injury. Spinal Cord August 2005;43(8):453–8.

[14] Smalheiser NR, Lugli G. Preparation of synaptosomes from postmortem human prefrontal cortex. bioRxiv January 1, 2014:006510.

[15] Simmons JP, Nelson LD, Simonsohn U. False-positive psychology: undisclosed flexibility in data collection and analysis allows presenting anything as significant. Psychol Sci November 2011;22(11):1359–66.

[16] Landis SC, Amara SG, Asadullah K, Austin CP, Blumenstein R, Bradley EW, Crystal RG, Darnell RB, Ferrante RJ, Fillit H, Finkelstein R, Fisher M, Gendelman HE, Golub RM, Goudreau JL, Gross RA, Gubitz AK, Hesterlee SE, Howells DW, Huguenard J, Kelner K, Koroshetz W, Krainc D, Lazic SE, Levine MS, Macleod MR, McCall JM, Moxley 3rd RT, Narasimhan K, Noble LJ, Perrin S, Porter JD, Steward O, Unger E, Utz U, Silberberg SD. A call for transparent reporting to optimize the predictive value of preclinical research. Nature October 11, 2012;490(7419):187–91. http://dx.doi.org/10.1038/nature11556.

[17] Rougier NP, Droettboom M, Bourne PE. Ten simple rules for better figures. PLoS Comput Biol September 11, 2014;10(9):e1003833. http://dx.doi.org/10.1371/journal.pcbi.1003833.

[18] Head ML, Holman L, Lanfear R, Kahn AT, Jennions MD. The extent and consequences of p-hacking in science. PLoS Biol March 13, 2015;13(3):e1002106. http://dx.doi.org/10.1371/journal.pbio.1002106.

[19] Darwin C. On the origin of species. London: John Murray; 1859.

D. MAKE YOUR DATA GO FARTHER

Even mice know the importance of proper randomization and are only too glad to help.

CHAPTER

15

Data Sharing and Reuse

DATA SHARING—WHEN, WHY, WITH WHOM

When one publishes an article, it is expected that the corresponding author will make the underlying data available to any reader. The editorial policy of *PNAS* is typical:

> To allow others to replicate and build on work published in PNAS, authors must make materials, data, and associated protocols, including code and scripts, available to readers. Authors must disclose upon submission of the manuscript any restrictions on the availability of materials or information. Authors must include a data availability statement in the methods section describing how readers will be able to access the data, associated protocols, code, and materials in the paper.... Authors are encouraged to use supporting information (SI) to show all necessary data or to deposit as much of their data as possible in community-endorsed publicly accessible databases, and when possible follow the guidelines of the Joint Declaration of Data Citation Principles. Research data sets should be cited in the references if they have a digital object identifier (DOI).... Such deposition may facilitate access to data during the review process and postpublication. In rare cases where subject-specific repositories are not available, authors may use figshare or Dryad...... Authors must make **Unique Materials** (e.g., cloned DNAs; antibodies; bacterial, animal, or plant cells; viruses; and algorithms and computer codes) promptly available on request by qualified researchers for their own use. Failure to comply will preclude future publication in the journal. It is reasonable for authors to charge a modest amount to cover the cost of preparing and shipping the requested material. Contact pnas@nas.edu if you have difficulty obtaining materials [1].

Note that there is a distinction between the data associated with a published article, which any reader can expect to receive, and unique materials generated by the authors, which are only pledged to be distributed to "qualified researchers for their own use."

It is one thing to state lofty principles for others to follow and another to follow them:

- "Give up my hard-earned data that took me 5 years and $5 million to acquire and another 3 years to analyze?
- To any reader? For any reason?
- That includes sending the data to my direct competitors who want to scoop me?
- To gadflies who mock me on their blogs and criticize my conclusions without even reading the paper carefully?

FIGURE 15.1   **In some fields, data sharing is not the norm.**

- To people I never heard of, who will probably misunderstand the data and handle it improperly?
- Even to companies that want to file a patent on my ideas if I do not file first?"

What do you think the response rate is when you ask a scientist for the data associated with a published article? The norms vary enormously from one community of scientists to another (Fig. 15.1 vs. Fig. 15.2), but data that are not publicly posted, but said to be "available on request," are rarely made available [2–4]. Nevertheless, allowing readers to see the data underlying a study, and to reanalyze them at will, is a big step forward in fostering research reproducibility. Fortunately, attitudes about data sharing are changing rapidly, and the infra-structure to permit data sharing is emerging as well.

DATA SHARING IS GOOD FOR YOU (REALLY)

As someone who has personally given out data to competitors and detractors alike, I believe that the reasons for *not* sharing data are mostly based on fear, and the risks are more theoretical than real. Most of the time when I have given out data, the recipients did not actually have the time, the attention span, the computer savvy, and/or the manpower to analyze the data sets! And there is far more glory in announcing a new experimental

FIGURE 15.2   That's more like it!.

discovery than in reanalyzing old data. Investigators are more likely to regard other people's data as garbage than as prize booty.

Besides, consider the upsides of sharing data (and reagents), which are as follows:

- Your article will be cited more, potentially a lot more.
- You will have established yourself as an open, collaborative white-hat type of person, and folks from around the world may be more likely to contact you for all sorts of initiatives and opportunities.
- Your research has demonstrated real-world impact and utility—others wanting to use your findings is a much higher bar than simply being cited or tweeted about. This is helpful when attempting to get or renew grants, get promoted, or establish commercial licensing.
- You will be contributing (at least indirectly) to new discoveries made by the scientists who use your data.
- If your data are combined with others, you may have access to the larger combined data and may contribute directly to making new discoveries.
- Moving your data or unique materials (such as antibodies or mouse strains) to public or even commercial repositories allows you to maintain these resources through funding ups and downs, laboratory moves, computer crashes, and all of the unforeseen events that happen over time. In my own case, we commercially licensed a variety of antibodies that we made, so that when other laboratories requested them, we did not need to maintain, test, store, and ship them out ourselves.

Data, like your children, take on a life of their own over time. My colleague Vetle Torvik and I developed a data set 10 years ago called "Author-ity," which uses a statistical model to predict which individuals have authored which articles in PubMed [5]. Our original reason for undertaking this work was to understand how new knowledge can emerge when different fields of study are brought together. We wanted to relate different fields to each other according to how they share the same investigators, are linked by investigators in each field who have worked directly with each other, or are linked even more indirectly through other colleagues. This somewhat arcane purpose was soon eclipsed by a flood of social scientists, science policy analysts, and economists who study innovation in medicine and who needed a clear idea of which individuals wrote what articles to carry out their own studies. We have periodically updated the Author-ity data set and given it out freely to academics around the globe, as well as licensed it to the National Library of Medicine (in the United States) for use as a prototype for them to build their own author tracking system. Lee Fleming's group created a different data set that assigned patents to individuals [6], which could be linked to our data set that assigns publications to individuals. By linking the two sets by common individuals, it is possible to see relationships between patents and publications, which are not otherwise possible. The original work was supported by a 2-year NIH grant, followed by two NSF grants (to connect the author publications to their patents and grant data). An ongoing 5-year NIH grant will improve the statistical model, further update and expand the data set, and provide the data publicly in a variety of formats for different types of users. A mighty oak has grown from sharing our original little acorn. And, nobody ever scooped us or made us regret handing out the data set.

Another example of how data sharing can lead to new knowledge involves linking data sets as an approach to predicting new uses for existing drugs: For example, one data set might be a list of drugs and the genes that they affect (up or down) in various organs. The other could be a list of diseases and the genes that are affected (up or down) in that disease, in various organs. If an approved drug affects genes altered in a disease, in the opposite direction from that produced by the disease itself, such a drug might be regarded as a new candidate treatment.

Should You Share Unpublished Data?

As strongly as I favor sharing published data, I am against sharing data prior to publication. This is not from fear of being scooped. Rather, I fear that I have not fully finished optimizing and double-checking the data cleansing steps, so that there may be frank errors.

Tip: Do not share data until the paper has not only been submitted but also has passed the peer-review stage and is in final preprint form. More than once have I submitted papers, only to have reviewers notice errors in the spreadsheets. This was an embarrassment when going through the publication process but would cause consternation and loss of reputation if I had given it out to another group.

DATA ARCHIVING AND SHARING INFRASTRUCTURE

Now that we have good intentions to share our data, we need a place to post or store the data that is permanent, secure, and has a **Digital Object Identifier (DOI)**, which uniquely identifies it regardless of its physical location(s). This requires infrastructure to create and maintain data repositories, together with efficient ways of describing the data (**indexing** and **metadata**), and ways to allow search engines to retrieve the right data efficiently and intelligently.

Each article (or other digital object) has metadata that describe what it is. For example, a published article will have metadata giving its title, the name of the authors, the name of the journal, major topics, and so on (see Chapter 16). The usefulness of metadata is greatly enhanced if they are formatted in a standard way that includes standard lists of attributes, and draws from concepts that are shared across all objects. Besides these metadata, each article may be indexed using features taken from the content (e.g., keywords that reflect major themes in the full text). The indexing helps to summarize the content and is added to the metadata, which makes it easier for search engines to find relevant articles. More important than mere words, however, are the underlying meaning of the words. An article might mention boat, ship, and Chinese junk as separate terms, which all map to the same general concept. Thus, articles may be indexed according to key concepts that are mentioned. Even further, the full text of an article may be processed to identify both concepts and relations between concepts, which can be written as a series of **triples** ("A" "relates to" "B") in what is known as **RDF** format. Such indexing is not intended for humans to read but rather allows computers to search and communicate in a more intelligent manner, a goal known as establishing the **Semantic Web**.

TERMINOLOGIES

Vocabularies, terminologies, and **ontologies** are all key concepts in managing data [7]. They are not always used in a consistent manner, and not everyone agrees on their definitions and scope, but roughly, **a vocabulary is simply a list of words found in a text collection**. The list may be ordered in different ways (e.g., alphabetically, or by frequency of occurrence, or by part of speech), but the words are not necessarily related to each other in any meaningful way.

In contrast, **a terminology is a list of words (or concepts) that form a hierarchy of relations between the members**. For example, brain > hindbrain > cerebellum > Purkinje cell, where each term is **a part of** the term on its left. This allows a computer to form certain limited inferences; for example, if I want to retrieve articles that discuss the brain, a search engine can infer that any article mentioning "cerebellum" will be relevant, even if the word "brain" does not appear.

In the field of medicine, several different popular terminologies are employed, each for a different purpose. Some of them are listed as follows:

1. The Medical Subject Headings (MeSH) hierarchy (https://www.ncbi.nlm.nih.gov/mesh) is used to assign terms to index biomedical articles in MEDLINE according to their major topics; the terminology is especially though not exclusively focused on diseases.
2. The Unified Medical Language System (UMLS; https://www.nlm.nih.gov/research/umls/) draws from a collection of medical terminologies including SNOMED CT, which

describes patient care encounters, and ICD-10, which is used to describe diseases and procedures for billing.

3. The UMLS Metathesaurus is a terminology that encompasses patient care, health services billing, public health, indexing, and cataloging of biomedical literature, as well as basic, clinical, and health services research. The terms consist of concepts that are assigned ID numbers, are related to other concepts, and are classified according to semantic type (e.g., medical device or clinical drug). Each ID number also covers a variety of synonyms and variant spellings that refer to the same underlying concept. Use of ID numbers fosters precision of writing: Does "cortex" in this instance refer to cerebral cortex, cerebellar cortex, or adrenal cortex?

4. Terminologies are common in specific fields of science, e.g., NeuroLex (http://neurolex.org/) is a terminology for classifying neuronal and glial cell types, channels and receptors, types of responses, etc. NeuroNames is a resource vocabulary for neuroanatomical terms [8].

ONTOLOGIES

Briefly, an ontology is a vocabulary in which the relations between the terms is described in much more detail than a simple hierarchy. **An ontology can be thought of as a schema that describes an event or situation fully—listing each of the components and their relationships to each other in context—to the point that a computer can use this information to make nontrivial inferences.**

Think of the schema that describes "Breakfast": It occurs in the morning, includes some kind of beverage, and includes some kind of solid food. Generally, breakfast is not the biggest meal of the entire day. Such schema covers a diversity of cases:

- A Japanese breakfast might consist of natto, raw egg, miso soup, rice, pickles, seaweed, and fish.
- An Israeli breakfast often has soft cheese, pickles, cucumber/tomato salad, and smoked fish.
- American breakfast generally has eggs, toast and/or pancakes, juice, and some kind of meat (bacon or sausage).
- A Continental breakfast may simply consist of a pastry, juice, and coffee or tea.

Such a schema suffices for a computer to make some inferences. For example, whenever breakfast is mentioned, it can be inferred that the event takes place in the morning. If smoked fish is mentioned, it is likely that the breakfast is NOT a typical American breakfast, and so on. (Note that the inferences are not guaranteed to be correct, just the most likely prediction given the data.)

The game of chess is nothing less than an ontology or schema. The game is fully defined by rules that describe the board, the pieces, the allowable moves, and the endpoints (when the king is subject to checkmate or when no legal move can be made). Thus, an ontology formalizes certain expectations; describes the entities and their usual ordering or arrangement; and specifies their relationship to other events or items.

Components of an Ontology

Ideally, an ontology is as complete as possible and consists of the following:

Individuals: instances or objects (the basic or "ground-level" objects)
Classes: sets, collections, concepts, types of objects, or kinds of things
Attributes: aspects, properties, features, characteristics, or parameters that objects (and classes) can have
Relations: ways in which classes and individuals can be related to one another
Function terms: complex structures formed from certain relations that can be used in place of an individual term in a statement
Restrictions: formally stated descriptions of what must be true for some assertion to be accepted as input
Rules: statements in the form of an if—then (antecedent—consequent) sentence that describes the logical inferences that can be drawn from an assertion in a particular form
Axioms: assertions (including rules) in a logical form that together comprise the overall theory that the ontology describes in its domain of application
Events: the changing of attributes or relations [9].

Gene Ontology

Perhaps the most famous and widely used ontology in biology is the gene ontology (GO) [10], which was launched in 1998 and originally described several model organisms (mouse, fruit fly, and brewer's yeast), the name of annotated genes, their molecular functions, subcellular localizations, and the biological processes they are involved in:

Molecular function is the function that a protein performs on its direct molecular targets; e.g., the insulin receptor has transmembrane receptor tyrosine protein kinase activity (GO:0004714), which means it catalyzes the reaction that adds a phosphate group to a tyrosine in another protein (its target).
Cellular component is the location where the protein performs its molecular function; e.g., the insulin receptor is located in the plasma membrane (GO:0005886).
Biological process covers the biological systems to which a protein contributes; e.g., the insulin receptor is involved in regulation of carbohydrate metabolic process (GO:0006109).

The GO has been expanded to include a variety of species and a variety of gene products (including not only genes that code for proteins but also noncoding RNAs as well). It should be noted that GO is something less than a true ontology, since it is not comprehensive, not entirely self-consistent, and does not encompass an entire schema or theory. Nevertheless, its popularity and usefulness cannot be denied. Perhaps the most popular use for GO is for analyzing trends that are shared over a large set of genes or proteins. For example, suppose an investigator finds that 100 genes are upregulated in response to a particular treatment; GO analysis might find that these genes tend to be involved in a shared common pathway such as regulating cell division.

Lots of Ontologies!

The National Center for Biomedical Ontology currently lists 538 different ontologies on its Web portal (http://bioportal.bioontology.org/). Another portal for openly available ontologies is http://www.obofoundry.org/. These run the gamut literally from soup to nuts. (Actually, soup and nuts may be the only two things for which ontologies do not exist!) For example, the Diabetes Mellitus Diagnosis Ontology (DDO) is an ontology for diagnosis of diabetes, containing diabetes-related complications, symptoms, drugs, lab tests, etc. The Adverse Event Reporting Ontology (AERO) is aimed at increasing quality and accuracy of reported adverse clinical events. The Venom Ontology (CU-VO) describes venoms and venom-related compounds. There are ontologies for drug names, for describing spinal cord regeneration, and for emotions, among others. Not all ontologies are open and public; for example, Ingenuity software is sold for analyzing biological pathways based on a proprietary ontology and knowledge base.

How Do Ontologies Assist With Data Sharing and Reuse?

A spreadsheet filled with numbers is worthless without knowing the details of the experiment, as well as a clear key explaining the **schema**—spelling out what the rows and columns mean. Consider this example: You are looking for articles on the clinical disease known as depression. You want to be able to distinguish depression the clinical disease (which has certain formal diagnostic criteria) from depression the mood (feeling down is not always a sign of depression, and not all who suffer from clinical depression feel down), from depression of the tongue, long-term depression of waveforms in the brain, and so on. Manic depression is a clinical disease too but not the same one as depression. Hospital notes and electronic health records need standardized terms and concepts that distinguish each of these. It would be crazy if Cerner (a big vendor of electronic health record systems) used the term "type I diabetes" to mean the same thing as "type II diabetes" does in Epic systems (another big vendor). Not only do ontologies make it much easier to mine the records within one hospital system, but they also allow those records to be shareable with other hospitals and to combine the data to mine for large-scale trends and patterns.

YOUR EXPERIMENT IS NOT JUST FOR YOU! OR IS IT?

Scientists traditionally value simplicity. Occam's razor is an oft-quoted principle that is variously translated as: "Among competing hypotheses, the one with the fewest assumptions should be selected" or "Entities should not be multiplied unnecessarily." In other words, keep your hypotheses as simple as possible. Similarly, when carrying out an experiment to test a hypothesis, a scientist is exhorted to collect all the data needed to get a clear yes/no answer—and nothing else. This frugality is fine in a world where each experimenter acts independently. But when data from an experiment are shared among different investigators, and as they are more and more being reanalyzed, combined, and otherwise reused in the future, then investigators need to have a different perspective:

1. One experiment often has more than one hypothesis; there is the original hypothesis you have before doing the experiment and the revised view of the same hypothesis you have afterward! **What you learn during the experiment may not simply test the hypothesis so much as alter, transform, and reformulate the hypothesis.** For example, recall our simple experiment in Chapter 1, where we hypothesized that social work majors may be more likely to be organ donors than economics majors. As researchers start surveying students, however, they may notice other differences between the two groups that might seem to be more informative or influential in terms of governing their altruistic behavior. Perhaps one group is disproportionately wearing crosses around their necks or displaying political bumper stickers. Regardless of whether the original hypothesis turns out to be supported or not, our perspective on how majors correlate with and reflect other personal choices will be changed. We may realize the value of obtaining ancillary data on the subjects (such as gender, age, religious observance, etc.), which may actually turn out to be more important explanatory factors than their majors.
2. Your experiments are more valuable when they can be directly compared with others. If I ascertain whether students are organ donors by checking the state organ donor registry, and everyone else in the field uses a different method, it will be hard to compare, much less combine, data across studies. A much better approach is to add **"glue" data**, that is, perform your measures BOTH ways. That way you can compare the results of the two measures directly on your own data and can have common ground for comparing to other studies.
3. You should always be mindful that data that you collect for one purpose may have value for future use by others for unanticipated purposes. For example, newborn blood is collected routinely for metabolic screening for the disease phenylketonuria (PKU) and then discarded. However, these samples could potentially provide a universal genetic registry, could be used for genetic screening of other kinds, and might be studied for all kinds of reasons, if only they were archived in a form that would permit reuse.

WHAT DATA TO SHARE?

Data sharing policies are designed minimally and defensively by journals—that is, they specify that authors should provide enough data to allow readers to check the validity of the findings reported in the article itself. This usually means attaching the final cleansed data (after removing errors and outliers, normalizing and transforming the values, binning, etc.). As we know, however, the cleansing process can influence the size, the significance, and even the direction of effects, so both the raw data and the cleansed data should be provided. And, of course, having the raw data are essential for another team to reanalyze the experiment or combine the data across studies.

What About Data on Individuals?

When publications report data that involve human subjects, or analyze information arising from clinical notes or electronic health records, the rules get complicated. Generally, to write a case report about a patient that can be identified from the information given in the article, one needs to have a signed consent form from the patient or their legal guardian or next of kin, even if the patient is currently deceased. The main question is whether the patients can be identified, and this can be surprisingly tricky. If I write a clinical case report about a 102-year-old man, that fact alone may make him identifiable, since there are so few centenarians alive. Similarly, a red-headed woman with six children living in a small town in Wyoming might be identifiable even without mentioning names. Or, if they have an unusual type of bone breakage or a rare disease, that might make them identifiable.

Moreover, even without the obvious identifiers (such as Social Security numbers, addresses, phone numbers, or emails), just knowing the DNA sequence of an individual may be enough to ascertain his/her last name (by comparison to sequences listed in public genealogy websites), and if one knows their age and the state in which they live, that may be enough to infer who the person is [11]. As a rule, then, clinical information at an individual level tends to be kept privately. Such data may be released on request, only after the requestors have shown that they have a legitimate need; are taking proper precautions to preserve privacy; and have approval for their proposed study from an **Institutional Review Board (IRB)**. Otherwise, the data may be shared publicly in "deidentified" form. Deidentification may strip the data of their usefulness for certain further analyses.

Proprietary data refer to data that are either owned by commercial companies or which involve trade secrets (like the formula for Coca-Cola). Pharmaceutical companies may simply choose not to publish studies on proprietary compounds. I have seen articles in which an investigator reports the effects of an agent that is described only by a company code, e.g., SP90Y, without giving its full structure. Most journals permit proprietary data to remain private and do not compel the authors to share with readers, although I think it is somewhat bad form to seek priority through publication without sharing the data in a form that others can use.

National security interests are another tricky subject. It is unlikely that someone will try to publish classified state secrets in an academic journal. But suppose a team is studying factors that control virulence of some infectious agent and has identified gene sequences that create superbugs: That information might be used for bioterrorism. Articles on computer security are also potentially sensitive. And there are no clear guidelines on what might be considered sensitive for national security. A paper on nesting sites for cranes might include photographs of areas near military bases that might potentially make someone nervous! Generally, academics are free to choose what to publish without having someone (even their funders) review and approve the article before submission to a journal. However, if you work in an area that has the potential to be sensitive (Fig. 15.3), I suggest raising the possibility with the journal editor at the time of submission.

FIGURE 15.3   **Research on zombies falls in a grey zone between human subjects and national security.** Best to consult the journal editor before submitting your article.

WHERE TO SHARE DATA?

Traditionally, data are displayed only in summary form within the figures and tables of the published article. The cleansed data might be attached to the article as a supplementary file (e.g., as one or more spreadsheets), together with a README file that explains how to read and use the data. Certain types of raw data such as microarray data, fMRI imaging data, crystallographic coordinates, and DNA and RNA deep sequencing data sets are appropriate for deposition into public domain-specific repositories (see below), in which case the authors generally provide the accession numbers in the manuscript.

Perhaps the most common way that authors currently share data is to post on their own project websites and provide the links (URLs) within the article. What could go wrong? First, as we warned in Chapter 14, unlike DOIs, URLs are not very permanent and are at high risk of being lost or modified. This is especially true when linking to websites that are not under your direct control. However, it is still a bad idea to link data to a personal URL even when they are under your own control! Personal and project websites are notoriously poor at backing up data, and URLs may change or data may be lost when project funding periods are over, when hard drives crash, or when personnel move to new institutions. Second, the data linked to the article should be *exactly* what was analyzed in the article. But nothing prevents the authors from adding, correcting, or indeed removing the data from their website. There is no enforcement of **version control**.

It is better to post data in a free, public repository such as Dryad, Github, or figshare, which has permanent DOIs and offers version control. However, it may not be clear to the reader who accesses the file which version is the original from the time of submission. And, ultimately, nothing prevents a file from being deliberately removed. For all these reasons, attaching the data directly to the article as a supplemental file is the best way to go, as long as the journal can handle the size and format requirements of the data set.

DATA REPOSITORIES AND DATABASES

Literally every discipline has public data repositories, often sponsored by governmental agencies or nonprofit organizations, which are highly utilized by specific communities of researchers. For example, NCBI Entrez suite (https://www.ncbi.nlm.nih.gov/) contains both curated databases and user-submitted data repositories with an emphasis on proteins and genes. The European Bioinformatics Institute (http://www.ebi.ac.uk/) has a similar suite. There are many independently hosted specialized databases as well: One of the oldest and most valuable databases is the Protein Data Bank archive (PDB) (http://www.wwpdb.org/), which houses 3D structural information. Biosharing.org is a portal of both databases and data repositories in the biological sciences (https://biosharing.org/databases/), whereas DataONE is a portal to data repositories on earth and the environment (https://www.dataone.org/). The US government has a portal to access its databases (https://www.data.gov). Other portals to data repositories across all sciences can be found at http://www.re3data.org/ and http://oad.simmons.edu/oadwiki/Data_repositories.

Apart from research data per se, there are also many open repositories for computer code and software, among them SourceForge, Github, CRAN and Bioconductor (for packages written in R), CPAN (for code written in Perl), and PyPI (for code written in Python). Finally, institutional repositories (for example, my university hosts INDIGO https://indigo.uic.edu/) archive intellectual products for their students and faculty, which may include published articles as well as manuscripts and data sets (although they may not be set up to handle large data objects).

SERVERS AND WORKFLOWS

Servers are another class of websites that are probably even more heavily visited and used than the official databases. These take information given by the user, process it using internal (possibly proprietary) algorithms, and return the predictions or analyses back to the user. A very highly utilized server is the BLAST tool (https://blast.ncbi.nlm.nih.gov/Blast.cgi), which takes nucleotide and protein sequences and finds the most similar sequences within a user-specified domain (e.g., human chromosomal DNA) (Fig. 15.4). A list of some of the better bioinformatics servers is maintained by the journal Nucleic Acids Research (http://www.oxfordjournals.org/our_journals/nar/webserver/c/). Other servers range from the mundane to the arcane—the PrimerQuest tool (https://www.idtdna.com/PrimerQuest/Home/Index) takes a nucleic acid sequence and predicts the best primers; SecretomeP (http://www.cbs.dtu.dk/services/SecretomeP/) takes a protein sequence and predicts how likely it is to be secreted from cells in a nonclassical manner. Servers give outputs that are usually based on published research but are not necessarily validated or optimized or better than other servers that use competing algorithms. So, buyer beware.

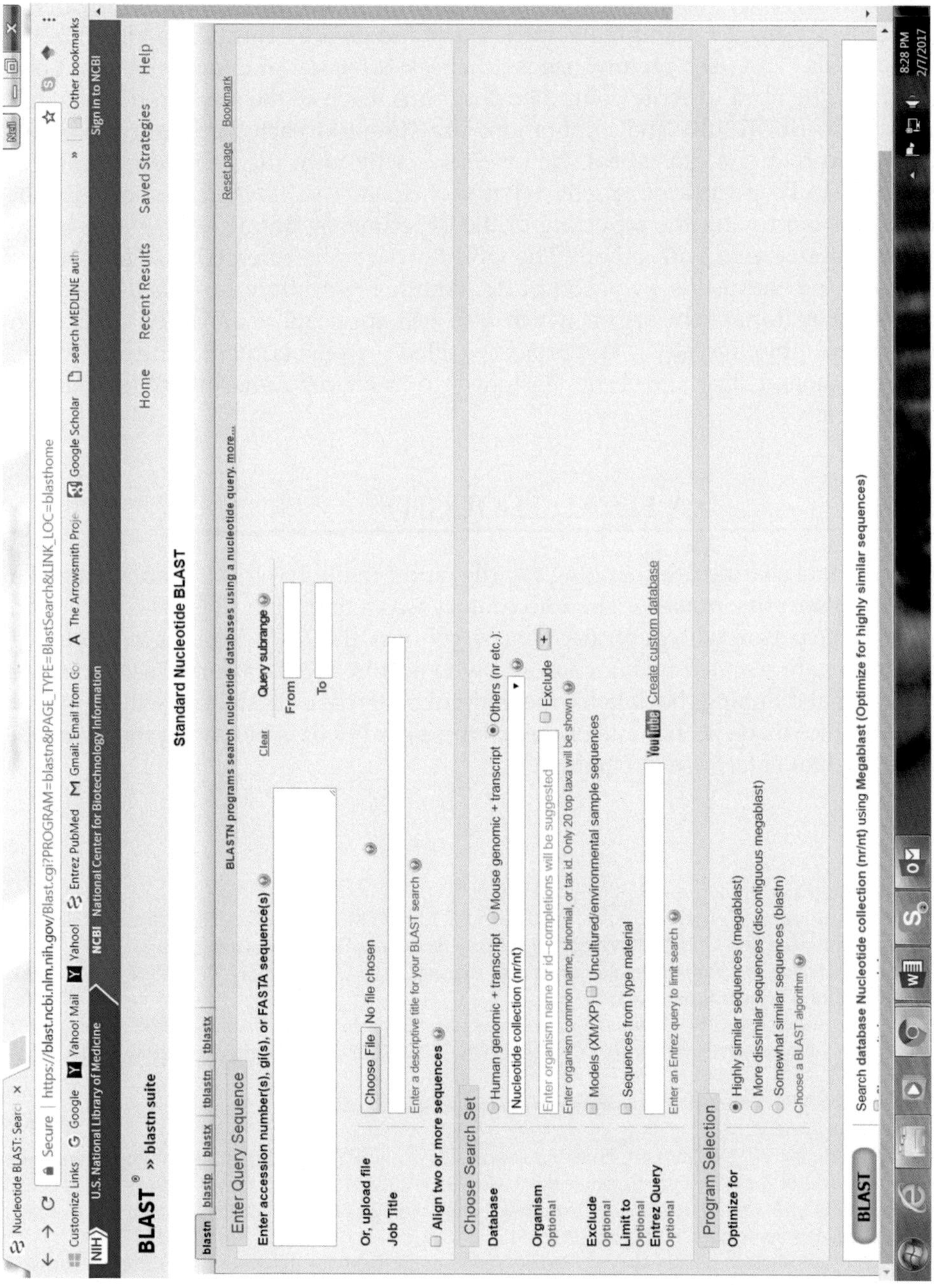

FIGURE 15.4 A screenshot of the Nucleotide BLAST server. *https://blast.ncbi.nlm.nih.gov/Blast.cgi.*

D. MAKE YOUR DATA GO FARTHER

Finally, there is a class of software called a **workflow,** which combines features of a data repository with those of a Web server and is explicitly intended to improve experimental reproducibility. The most well-known workflow is Galaxy (https://usegalaxy.org/), which allows users to manipulate nucleic acid sequences through a pipeline, to carry out an analysis. The user merely needs to check off tasks and parameter options from a list, without the need to write computer programs. Each of the steps in the pipeline is annotated, the interim results and annotations are stored at each step, and the entire pipeline is stored so that one can repeat the process exactly with the next new sequence. This makes it easy to keep track of exactly what was done. This is valuable not only for saving time and for improving the reporting of the experiments, but also the entire workflow can be saved and shared with others. The workflows can be accessed on Web servers or installed locally on the user's own computer. Another workflow is Apache Taverna (https://taverna.incubator.apache.org/), which also has applications in other fields such as astronomy and biodiversity. A project called myExperiment (http://www.myexperiment.org/home) has a website designed to encourage sharing of workflows among investigators.

A FINAL THOUGHT

Attitudes about data sharing are changing rapidly, and I think that this is emblematic of a larger global shift in scientific practice. The 20th century was a time of individual competition and credit went to the person who published first. In contrast, the 21st century is an era of big science: Experiments are planned by big consortia who acquire big data, and discoveries are made by combining and mining the data for meaningful patterns. Data sharing will no longer be an individual choice made at the time of publishing a study but is becoming the glue that drives the experimental enterprise altogether.

References

[1] http://www.pnas.org/site/authors/editorialpolicies.xhtml.
[2] Tenopir C, Allard S, Douglass K, Aydinoglu AU, Wu L, Read E, Manoff M, Frame M. Data sharing by scientists: practices and perceptions. PLoS One 2011;6(6):e21101. http://dx.doi.org/10.1371/journal.pone.0021101.
[3] Fecher B, Friesike S, Hebing M. What drives academic data sharing? PLoS One February 25, 2015;10(2):e0118053. http://dx.doi.org/10.1371/journal.pone.0118053.
[4] Milia N, Congiu A, Anagnostou P, Montinaro F, Capocasa M, Sanna E, Destro Bisol G. Mine, yours, ours? Sharing data on human genetic variation. PLoS One 2012;7(6):e37552. http://dx.doi.org/10.1371/journal.pone.0037552.
[5] Torvik VI, Smalheiser NR. Author name disambiguation in MEDLINE. ACM Trans Knowl Discov Data July 1, 2009;(3):3, 11.
[6] Li GC, Lai R, D'Amour A, Doolin DM, Sun Y, Torvik VI, Amy ZY, Fleming L. Disambiguation and co-authorship networks of the US patent inventor database (1975—2010). Res Policy July 31, 2014;43(6):941—55.
[7] Spasic I, Ananiadou S, McNaught J, Kumar A. Text mining and ontologies in biomedicine: making sense of raw text. Brief Bioinformatics September 1, 2005;6(3):239—51.
[8] Bowden DM, Song E, Kosheleva J, Dubach MF. NeuroNames: an ontology for the BrainInfo portal to neuroscience on the web. Neuroinformatics 2012;10(1):97—114. http://dx.doi.org/10.1007/s12021-011-9128-8.
[9] https://en.wikipedia.org/wiki/Ontology_(information_science).

[10] Ashburner M, Ball CA, Blake JA, Botstein D, Butler H, Cherry JM, Davis AP, Dolinski K, Dwight SS, Eppig JT, Harris MA, Hill DP, Issel-Tarver L, Kasarskis A, Lewis S, Matese JC, Richardson JE, Ringwald M, Rubin GM, Sherlock G. Gene ontology: tool for the unification of biology. The Gene Ontology Consortium. Nat Genet May 2000;25(1):25–9.

[11] Gymrek M, McGuire AL, Golan D, Halperin E, Erlich Y. Identifying personal genomes by surname inference. Science January 18, 2013;339(6117):321–4. http://dx.doi.org/10.1126/science.1229566.

D. MAKE YOUR DATA GO FARTHER

Data sharing: All the cool kids are doing it!

16

The Revolution in Scientific Publishing

JOURNALS AS AN ECOSYSTEM

The end goal of most research is to be published, and so it is appropriate for us to consider the ways in which journals can either hinder or facilitate the quality of research.

The earliest scientific journals date to 1665, when the English *Philosophical Transactions of the Royal Society* and French *Journal des sçavans* first published research findings. Most journals nowadays are hosted by commercial publishers; others are published by scientific societies, universities, and even funding agencies. In all cases, however, a journal is intended to fill some need and niche in the scientific community. When new emerging topics appear (such as nanotechnology or robotics), new journals appear. Some journals specialize in particular types of articles such as clinical case reports, essays, hypotheses, or descriptions of software, which do not fit the publication criteria of journals that focus primarily on experimental research. Some journals aim for newsworthy findings, some for large definitive studies, and others for rapid communications of preliminary results. Some journals are international in scope and others focus on regional issues.

As new media technologies have appeared, and as scientists are focusing more and more on research robustness and reproducibility, the publishing enterprise has responded, and we are in the middle of a revolution that is changing how scientific studies are reported and disseminated. This is very good news!

PEER REVIEW

Traditionally, all reputable journals send submissions through peer review before deciding whether to publish them. Briefly, when an author submits a manuscript to a journal, it is first screened by an editor for suitability (topic, appropriateness, and quality), then either rejected outright or sent for review. Generally the editor (or a field editor) chooses two to five reviewers to read the paper in detail. The reviewers may be known experts or peers who have published on similar topics, although sometimes researchers in different fields may be consulted for their perspective (e.g., statisticians or ethicists may be asked to weigh in on a

FIGURE 16.1   Could have been worse, I guess.

manuscript). Most commonly, the identity of the author is known, but the reviewer remains anonymous (single-blind review), although in some fields, it is common to do a double-blind review, in which case the manuscript is stripped of identifying information. Reviewers return their reviews to the field editor, who makes a provisional decision (accept, reject, minor revisions needed, or major revisions needed) and sends the decision and the reviews to the author (Fig. 16.1). The author may prepare and resubmit a revised manuscript, which may be rereviewed or acted on directly by the field editor. Traditionally, reviews remain confidential. If the author rejects the suggestions of the reviewers, he or she may appeal a decision to the chief editor. The majority of accepted manuscripts are revised at least once or twice.

Peer review is time-consuming and subjective. Reviewers can also be consciously or unconsciously biased against certain authors, genders, topics, or innovative ideas [1]. But it is the primary way that journals ensure the quality of the research articles that they publish.

JOURNALS THAT PUBLISH PRIMARY RESEARCH FINDINGS

High-Impact Journals

At the top of the heap are a small handful of "high-impact" journals: *Nature, Science, Cell,* and *PNAS* (Proceedings of the National Academy of Sciences, the United States), plus three medical journals, *New England Journal of Medicine, Lancet,* and *JAMA* (Journal of the American Medical Association). These are very large, prestigious, multidisciplinary journals that aim to publish only the most important and timely findings. To publish in one of these, an article must be newsworthy, relevant to multiple disciplines, appeal to the general public, report very large or surprising effects, or have clear political or policy implications.

D. MAKE YOUR DATA GO FARTHER

It is probably more accurate to consider the high-impact venues as magazines rather than scholarly journals. About 95% of all articles submitted to these journals are rejected (often by editors without being sent for review), which is not surprising considering that submissions will be competing for space with big science articles such as NASA's Jupiter flyby missions, the entire human genome draft sequence, and supercollider studies that involve hundreds of physicists.

Conventional wisdom has been that getting published in one of these journals is essential to establish your spot in science and is essential to getting promoted and funded. That belief has directly or indirectly shaped people's entire careers—determining what topics and methods they pursue, which institutions they choose for training, which mentors they choose, and the size of the laboratories they direct. In my opinion, publishing in a high-impact journal is somewhat like winning the Academy Award in movies: wonderful for your resume, certainly, but not essential. In fact, not only is there a long list of distinguished actors who have never won Oscars, but also is there a long list who have never even been nominated— Richard Gere, Meg Ryan, Alan Rickman, and Marilyn Monroe, to name a few. Similarly, there is a long list of articles that were rejected by high-impact journals and other leading journals of the day, yet subsequently won the Nobel Prize [2].

I remember sitting in a resort bar in Tuscany during a conference (Fig. 16.2) with the late E.G. "Ted" Jones, a very senior and accomplished neuroscientist. We were admiring the view over the hillsides, and he was introducing me to grappa, a spirit distilled from the dregs of the grape. In this pleasant venue, he confided that in his department, an assistant professor is

FIGURE 16.2 Attending conferences is another important part of a scientist's life, often more pleasant than writing articles.

TABLE 16.1   Publications of E.G. "Ted" Jones

	Journals	Frequency
1	*J Comp Neurol*	53
2	*Brain Res*	41
3	**J Neurosci**	37
4	*Exp Brain Res*	18
5	**Proc Natl Acad Sci U S A**	17
6	*Neuroscience*	15
7	*J Neurophysiol*	12
8	*Cereb Cortex*	8
9	*Philos Trans R Soc Lond B Biol Sci*	7
10	*J Neurocytol*	6
11	*Arch Gen Psychiatry*	5
12	*Eur J Neurosci*	5
13	*Biol Psychiatry*	4
14	*Brain Res Mol Brain Res*	3
15	*J Anat*	3
16	*Mol Psychiatry*	3
17	*Prog Brain Res*	3
18	*Adv Neurol*	2
19	*Annu Rev Neurosci*	2
20	*Brain*	2

Shown are the top 20 journals ranked by the number of articles published. Journals that he recommended for assistant professors to publish in are given in bold.

judged only by his or her publications in the high-impact journals plus *Neuron* and *Journal of Neuroscience*. Nothing else is of any consequence. He was being honest and candid, exaggerating only slightly, and giving me the benefit of his wisdom and experience. At the same time, the vast majority of his own publications had been in other, excellent but more specialized journals (Table 16.1). And when I look at other famous and productive professors, in a variety of fields, I see a similar story. They may have had one or a few articles in high-impact journals, particularly early in their careers, or maybe none at all. Most of their articles appear in specialty journals (first to third tiers; see below), and the common thread is that their body of work speaks for itself. Personally, I suspect that the primary role of high-impact journals as gatekeepers of high quality is diminishing—much as in the field of pop music, Justin Bieber could be discovered directly from his self-posted performances on Youtube without needing to secure a record contract and promotion.

Part of the Solution or Part of the Problem?

The revolution in scientific publishing is centered around principles such as free and open dissemination of findings and the reporting of complete details concerning an experiment, including posting the raw and cleansed data. Historically, the policies of the high-impact journals have worked against these principles. Because they are published primarily in print form, space is extremely tight, and articles are short and lacking full details. (As online supplements have become available, the online supplements may be many times longer than the print-based article itself.) The list of references is restricted in number, limiting the authors' ability to cite all relevant work in a scholarly manner. Authors are prohibited from having circulated the work in any form, including posting preprints online. And the copyright belongs to the journal and not to the author. Anyone wishing to read the article must pay either a subscription fee or a reprint fee. The high-impact journals recognize that research reproducibility and open science are big trends, and so they would like to be out in front leading the way—and yet they are also making a lot of money from fees. As of the writing of this book in 2017, these restrictions are being nibbled down but are by no means gone. So, high-impact journals are simultaneously part of the solution and part of the problem!

More subtly, the magazine-like aspects of high-impact journals encourage viewing science as a horse race (winners and losers), and they highlight awards, scientific celebrities, politics, and news. Although high-impact journals certainly do not condone cutting corners or outright fraud, the "publish-or-perish" mentality is encouraged that incentivizes such behavior.

Tiers of Scholarly Research Journals

First-Tier Scholarly Journals

The top tier of scholarly journals judge articles according to specific scientific criteria of novelty, importance, and rigor. They tend to cover specific disciplines broadly and historically have often been supported by scientific societies. Some examples in the biomedical domain include the *Journal of Biological Chemistry*, *Journal of Cell Biology*, *Journal of Neuroscience*, *PLoS Computational Biology*, or *eLife*. These are generally large journals and generally publish big, complete, definitive studies. Besides charging readers (or libraries) for subscriptions, often authors are expected to pay page charges as well. A few journals are a hybrid of first tier and high impact insofar as the editors assess newsworthiness and timeliness of articles in addition to inherent scientific quality. *Neuron*, *Immunity*, and *Nature Neuroscience* probably fit here.

Beyond the first tier, journals can often be ordered among themselves according to several dimensions. For example, journals can have broad versus narrow scope: If *Journal of Neuroscience* covers this discipline broadly, then *Journal of Neurochemistry* and *Journal of Neurophysiology* are narrower, and *Sleep and Sleep Disorders* are progressively narrower still. Journals can also be ordered according to their size (measured in number of articles, readership, or total citations) and geographical scope. *Journal of Immunology* is broader than the *Scandinavian Journal of Immunology*.

Second-Tier Journals

I will go out on a limb and list a few top scholarly journals (among hundreds) that have more specialized scope than the first tier: *Learning & Memory*, *Genome Research*, and *Nucleic Acids Research*. Journals in the second tier often publish a mix of studies including methods papers, short communications, and articles that employ bioinformatics as a primary method (i.e., instead of presenting new experimental data, they report analyses of existing public or published data).

Third-Tier Journals

Here I will include the majority of reputable journals, published by both mainstream and open access publishers, which publish "solid" mainstream research but where articles may be more preliminary, have more limited scope, or appeal to a subspecialty audience.

Fourth-Tier Journals

Finally, among high-quality scholarly journals, I will include those that reflect particular editorial slants or historical events. For example, the *Journal of Machine Learning Research* arose when 40 editors of the subscription-based journal *Machine Learning* resigned to form a new journal that would allow authors to retain copyright over their papers and to provide articles freely available on the Internet. The journal *Evolutionary Theory*, founded by Leigh Van Valen at University of Chicago, was a place to publish unconventional ideas or approaches online, rapidly and at low cost. The *Worm Runner's Digest* founded by James V. McConnell at University of Michigan published serious scientific studies mixed in with humorous pieces. My own journal, *DISCO (Journal of Biomedical Discovery and Collaboration)*, hosted by the University of Illinois Library System, http://journals.uic.edu/ojs/index.php/jbdc/, is devoted to studies of scientific practice and factors that influence scientific innovation. These fourth-tier journals tend to be small and have a definite personality that is linked to the chief editor.

The high-impact journals and first-tier journals are larger, have broader scope, often are better established, and often have greater prestige than the third- and fourth-tier journals. However, it is important to emphasize that experimental design, controls, data analysis, and reporting are not necessarily any better in the high-impact journals. Tiers do not correlate in any general way with the quality of the journal peer-review process, the reputation of the editorial board, or the resulting quality of the articles that they publish. **Breakthroughs and classics appear in all tiers**. The *Bulletin of Mathematical Biophysics*, edited by Nicholas Rashevsky (also at University of Chicago), was a fourth-tier journal that published McCulloch and Pitts' "A logical calculus of the ideas immanent in nervous activity" [3], which as we saw in Chapter 13, influenced the design of digital computers as well as launched the field of artificial intelligence.

Of course, beyond these tiers, there are a whole slew—hundreds, if not thousands—of journals that purport to review articles by peer review but have low quality, or even fall into a disreputable or junk category. One can find lists of "predatory" publishers and online journals that allegedly will publish any crap or nonsense that you send them, as long as you pay a fee. (Note that these lists were compiled by individuals by a not-totally-transparent process and should not be taken as definitive.) In parallel, there are also for-profit conferences, which will accept papers regardless of merit as long as the authors register for the conference. The existence of predatory publishers and journals is doubly unfortunate, for not only do they provide bad venues to present good research, but they also unfairly taint the reputations of high-quality journals that charge author acceptance fees, and/or open access fees (see below), yet maintain rigorous editorial standards and careful peer review.

INDEXING OF JOURNALS

How to avoid submitting your work to a junk or predatory journal? The short answer is to choose journals that are indexed by major indexing services, and especially choose journals in which articles appear that are important in your own field.

In medicine and related biomedical fields, one can trust journals that are indexed in **MEDLINE**. This is a curated collection of ~5600 established journals maintained by the National Library of Medicine (the United States), which takes into account not only the topical scope of the journal, but also various quality indicators related to peer-review policies, ethical guidelines, and how financial conflicts of interest are disclosed. MEDLINE includes journals from all tiers. One of the nice things about MEDLINE is the way that it is extensively indexed. Each article in MEDLINE is read in its entirety by a PhD-level curator who chooses an average of 8–20 **Medical Subject Headings (MeSH terms)** that summarize the major topics discussed therein. As well, articles are indexed according to publication type (e.g., Clinical Trial) and many other features that are presented in formatted fields. Together, the **MEDLINE record** for each article comprises **metadata** (i.e., data that describe data) that not only assist in finding the article when searching online but also provide a useful summary of its content. As of the start of 2017, MEDLINE indexes about 26 million articles.

PubMed Central (PMC) is also maintained by the National Library of Medicine but differs from MEDLINE in key respects. First, it is a collection of full-length articles, not merely their records. Second, it is quite heterogeneous, both in terms of its topical coverage and in terms of its sources. Historically, PMC was created in 2000 so that research that had been supported by grants from the National Institutes of Health (NIH) could be made freely available online to anyone, without the need to pay any subscriptions or fees. Any article that acknowledges NIH support is required to be deposited in PMC, regardless of the topic or which journal it was published in. Articles on art history, social work, electronics, or religion might be included in PMC, and even articles published in "junk" journals might be included. On the other hand, PMC also has a curated collection of journals that deposit all of their articles into PMC, regardless of whether they were supported by NIH. Most of these journals cover biomedical topics, but areas that are broadly relevant to medicine are included as well, e.g., crystallography, computer science, or information science. Finally, journals that publish articles according to a hybrid model (some authors pay fees so that their articles can be viewed by anyone; the remainder can only be viewed by subscription) may choose to deposit their open access articles into PMC. Subscription-based publishers are allowed to enforce a temporary embargo on accessing their articles in PMC, up to 2 years after publication. Overall, at the time of writing this, PMC contains over 4 million full-text articles and is growing rapidly.

PubMed is a search engine, hosted at https://www.ncbi.nlm.nih.gov/pubmed, which allows users to search MEDLINE (including the most recently published articles that have not been indexed by curators yet), PMC, and a variety of other specialized collections such as biomedical textbooks and government technical reports. I am a big fan of PubMed. It works hard to adjust its search interface to anticipate the needs of its diverse users. My own research team has created a search interface that carries out PubMed searches and then builds on them, allowing users to choose various tools that summarize and mine the articles in ways that go beyond the capabilities of PubMed. Our suite of tools is called Anne O'Tate and is freely accessible to anyone at http://arrowsmith.psych.uic.edu/cgi-bin/arrowsmith_uic/AnneOTate.cgi.

Another major curated collection of articles is **Embase**, which is broader than MEDLINE and covers medicine and biomedical science, with emphasis on pharmacology, toxicology, pharmaceutical science, and clinical research; it also has substantial coverage of allied health subjects such as nursing, veterinary science, and physical therapy. **CINAHL** is another collection specializing in nursing and allied health, whereas **PsycINFO** covers the psychological, behavioral, and social sciences.

Certain societies maintain a digital library of articles published under their sponsorship, notably ACM (Association for Computing Machinery) in computer science, IEEE (International Electronics and Electrical Engineers) in engineering, and ASIST (American Society for Information Science and Technology) in library and information science. Several noncommercial search engines are devoted to computer science and related subjects, including **dblp** (http://dblp.uni-trier.de/) and CiteSeerX (http://citeseerx.ist.psu.edu/index).

Perhaps the most popular search engine for finding articles, in all domains, is **Google Scholar (GS)** (https://scholar.google.com/) because of its speed and its wide coverage. Another advantage of GS is that it can help find full-text articles linked to publisher's sites. On the other hand, it is difficult if not impossible to download more than 1000 articles at a time, or resort and rerank articles according to different criteria, which are common tasks performed via PubMed or Anne O'Tate. Another problem with GS is that since it is hosted by a commercial outfit, it might be canceled or not maintained in the future. This is not a theoretical risk but a real possibility: note that a competing system, Microsoft Academic Search (https://academic.microsoft.com/) was completely decommissioned a few years ago and only recently was relaunched.

Finally, two popular commercial databases are **Web of Science** and **Scopus**. These are curated collections with very broad scope—science, mathematics, engineering, technology, health and medicine, social sciences, and arts and humanities. Because they are subscription based, they are mainly used at universities, and I personally find them to be clunky compared to some of the other choices. However, these are especially valuable if you are interested in tracing an article's **citations** (i.e., the articles that cite a given article, or are cited by it).

ONE JOURNAL IS A MEGA OUTLIER

In our discussion of journals so far, we have not mentioned *PLOS ONE*—one of the largest journals in the world, and one of the newest, having been established only in 2006. Arguably *PLOS ONE* is also the journal having the most important impact on scientific publishing today. Like Wikipedia, which transformed the way that the world accesses factual information, *PLOS ONE* is the online journal that has succeeded in transforming the way that scientists publish.

PLOS ONE publishes primary research articles (in any field of science), insists on technical rigor, and enforces good reporting practices (e.g., posting the raw, uncropped versions of photos used in figures; describing methods in full detail; ensuring that all data are available for reuse and reanalysis). At the same time, any sound work will be accepted regardless of whether the editors or reviewers feel that it has importance, general interest, or news value. Authors are charged fees for accepted articles but these are much less than similar fees charged by more traditional journals (\$~1000 vs. \$3000—\$4000). All published articles are freely available for users to view and make use of, and readers can post comments, which are attached publicly to the published article.

Whatever *PLOS ONE* did that went so right, it has a huge worldwide audience, and its articles include many that are well publicized (thanks, in part, to an active journal-based Twitter and press release campaign). Many authors who get their initial rejections from *Science* or *Nature* now go straight here, to avoid spending additional months or years fighting a succession of reviewers and editors in search of an eventual home. Indeed, to give one example that I know of personally, one of my colleagues had an article rejected from *Nature* without review—sent it to *PLOS ONE*—and subsequently saw his article profiled in *Nature* as a news item! As a result, *PLOS ONE* acts effectively both as a high-impact journal AND a third-tier journal.

In the face of this mega success, several other publishers have created mega journals of their own, notably *Scientific Reports* (covering biological and clinical sciences, recently overtaking *PLOS ONE* as the world's largest journal), *Sage Open* (for behavioral and social sciences), and *PeerJ* (with sections in both biomedical and computer sciences), among others. (Disclosure: I am a volunteer editor for both *PLOS ONE* and *Scientific Reports*.) I do not see any sign that mega journals will swallow up traditional first-tier journals. However, *PLOS ONE* in particular has made authors generally aware of, and accepting of, such practices as reporting checklists, data sharing, online supplementary files, and open access. In turn, this is putting pressure on traditional journals to provide these services to their authors.

WHAT IS OPEN ACCESS?

The short answer is that an open access article is one that can be read (full-text) by anyone, without the need to have a subscription or pay for a reprint.

However, there are two levels of open access. For example, Nature recently announced that it would allow **free viewing** of articles in ReadCube read-only format, which can be annotated **but not copied, printed, downloaded, or mined by computers**. In contrast, the Budapest Open Access Initiative defined open access as follows:

> By 'open access' to this literature, we mean its free availability on the public internet, permitting any users to read, download, copy, distribute, print, search, or link to the full texts of these articles, crawl them for indexing, pass them as data to software, or use them for any other lawful purpose, without financial, legal, or technical barriers other than those inseparable from gaining access to the internet itself. The only constraint on reproduction and distribution, and the only role for copyright in this domain, should be to give authors control over the integrity of their work and the right to be properly acknowledged and cited [4].

The Creative Commons licenses are the major licenses for distributing intellectual works in an open access manner (including scientific, artistic, and educational works). However, this is actually not a single license, but a family of related licenses, which has different levels of restrictions that authors can choose from:

- CC0 license donates the work to the public domain and waives all rights.
- CC-BY licenses allow the user to copy, distribute, display, and perform the work; and make derivative works and remixes as long as the original creator is given attribution.
- CC-BY-SA license says that the user may distribute derivative works (but cannot do so under a license that is more restrictive than the one applying to the original work).
- CC-BY-NC license says that the users can copy, distribute, display, and perform the work and make derivative works and remixes but only for noncommercial purposes.
- CC-BY-ND license prevents the user from modifying and distributing modified versions of the original work.
- A common license for distributing open source software is the GNU General Public License (https://www.gnu.org/licenses/gpl-3.0.en.html), which allows users to share and modify software freely (with attribution).

There are two types of open access publication venues, called green versus gold. **Green open access** refers to posting the published article somewhere online (other than in the journal itself). This may be the author's own website, or an institutional repository if you are affiliated with a university that has one. Generally, if the original article was published in a traditional (nonopen) journal that holds the copyright, the author is allowed merely to post a prepublication version of the accepted manuscript, and this may be further subject to a temporary embargo period. Green open access articles are easy for the author to post and do not incur article fees, but depending on how they are indexed and ranked by search engines, and whether the URL is maintained by the author, the copies may not be permanently posted or easy for people to find!

Green open access is related to, but distinct from, the posting of **preprints** online—that is, posting manuscripts either prior to or after submitting to a journal but before acceptance or publication. The majority of journals nowadays will allow authors to post preprints and still will consider the manuscript for publication (the primary exceptions being some high-impact journals). Posting of preprints may occur either on the author's own website or in an open access preprint repository such as arXiv or bioRxiv (see below).

In contrast, **gold open access** is when the journal directly publishes the article in open access format. Often (but not always), journals charge authors' fees for publishing open access articles. Journals may be completely open access, or they may publish according to a hybrid model comprising both subscription-access and open-access articles. (Potentially, a single article may find itself posted in both green and gold open access venues.)

Tip: Be aware that not every journal that offers "gold open access" actually offers open access in its full form. An author may pay a substantial fee for open access only to find out that readers can view their article but not download it, copy it, use figures from it, share it with students, add it to a database, etc.

Preprint Repositories

The first preprint repository, arXiv, was the brainchild of Paul Ginsparg, a physicist at Cornell University; over the past 25 years it has become the primary mode by which physicists publish articles and by which other physicists keep up with the advances in their fields. In other words, posting a preprint is now the primary means of publication in physics, with later publication in a peer-reviewed journal being considered secondary. The scope of arXiv covers physics, mathematics, and computer science, as well as computational biology and biophysics. One mathematician, Grigori Perelman, even won the million-dollar Millenium Prize for proving the Poincaré Conjecture and posting the solution on arXiv, without ever publishing in a peer-reviewed journal at all.

Biologists have shown relatively little willingness to post preprints online until recently, in part because of restrictions imposed by journals, and in part because biomedical findings can be repeated quickly and thus authors are afraid of being scooped (scientific priority is attached only to a peer-reviewed publication). Several preprint repositories launched in the 2000s by *Nature* (*Nature Precedings*) and *Genome Biology* failed. However, in 2013, buoyed by the success of *PLOS ONE* and the rapidly changing publication environment in biomedical sciences, Cold Spring Harbor Laboratory founded bioRxiv, which appears to have rapidly taken flight and gained mainstream acceptance. As I write this, ASAPbio is planning to create a unified preprint repository that will coordinate with arXiv, bioRxiv, and others.

Several innovative journals are being launched that blend the concepts of preprints and postprints into a single publication venue. For example, *F1000RESEARCH* is a journal in which submitted papers are immediately published, and then peer review takes place afterward (so that articles are either endorsed or not by reviewers). *PeerJ* publishes both preprints and peer-reviewed articles, although they are kept in separate sections. *PLoS Currents* is devoted to disseminating rapid communications on disasters, epidemics, and certain specific diseases such as Huntington's disease and muscular dystrophy. These articles are screened by experts for suitability but do not undergo formal peer review. Finally, *RIO Journal* publishes a variety of research-related items (e.g., research proposals, slide shows, etc.) and the type of peer review varies according to the type of item.

Prospective Registration

One of the ways that an investigator can guard against changing their outcomes or analyses after the fact is to publish the hypothesis and experimental design as a protocol paper, prior to starting the work. This is especially advisable for clinical trials. (Note: This is in addition to registering the trial itself in a clinical trial registry.) Systematic reviews (Chapter 10) also are amenable to detailed design protocols to ensure that the topic questions and inclusion criteria are unambiguous and that the types of acceptable evidence and sources are stated in advance. Several journals publish clinical trial and systematic review protocols routinely.

It has also been suggested that prospective registration of laboratory experiments would improve their robustness and reproducibility, and several journals now encourage the publication of results from registered experiments whether they prove to be positive or negative. The journal *Cortex* has launched a new type of article in 2014 called Registered Reports, in which an author submits the proposal of an experiment—and if accepted, the journal commits to publishing the results regardless of outcome. Several other journals have followed, including *eLife, Royal Society Open Science,* and *Comprehensive Results in Social Psychology.*

D. MAKE YOUR DATA GO FARTHER

This is a bold experiment in scientific practice, but I think it remains to be seen how popular it will be among laboratory investigators, who are used to tinkering and exploring as experiments progress.

IMPACT FACTORS AND OTHER METRICS

I am not a fan of ranking scientists, articles, journals, or universities. I understand the reasons why people find this interesting and important, of course, but we cannot rank the great. Jesus and Socrates wrote nothing at all. Gregor Mendel, founder of genetics, wrote a monograph on pea plants—a classic worthy of study by all biology students—that languished in obscurity for 35 years after its publication. Sigmund Freud was the most cited scientist of the 20th century: Does that imply that he was the greatest?

The earliest, and still the most influential, metric is the **impact factor (IF)** proposed by Eugene Garfield back in 1955 [5] and applied to journals. One looks back the preceding 2 years and makes a list of all the "substantial" articles published in that journal = N (e.g., including research articles and reviews but excluding editorials, news items, letters to the editor, etc.). Then, count all of the citations C made by all articles anywhere during the current year to the N journal articles published during the preceding 2 years. The impact factor (IF) = C/N. Roughly speaking, the IF says how many **citations per article** the journal has had in that recent 2-year window, on average.

High-impact journals have the highest IFs, which is not surprising since they choose articles, in part, because of their immediate impact and newsworthiness. Within a given field or discipline, first- and second-tier journals tend to have higher IFs than the third- and fourth-tiers, which is taken as validating the IF—although the same relative rankings would also hold if the tiers were simply ranked by number of articles published! There is nothing inherently wrong with using the IF—it has some value and is as good as any other single metric for ranking journals. This is like using one's weight as a single number to describe one's health: Not meaningless, but you need a lot of additional information for it to have meaning.

The problems with the IF are numerous, however, due to the fact that it is widely misunderstood and misused. One issue is that you cannot use the IF to compare different fields, because different disciplines have different citing behaviors and timeframes. One of the leading molecular biology journals, *Molecular Cell*, has IF = 13.96 (2015), whereas a top dentistry journal, *Periodontology 2000*, has IF = 4.95, and a top mathematics journal, *Annals of Mathematics*, has IF = 3.1. This difference reflects the different number of references per article in the two fields, the number of practitioners worldwide doing similar research topics, the relative pace of the fields, and the relative timeline for citation (do they tend to cite only extremely recent studies or older ones?). Articles in history and philosophy tend to have much longer timelines and often cite articles that are decades old. Also, one cannot use IFs to compare different types of journals. Review articles tend to have hundreds of citations and in turn are cited heavily relative to research articles; thus, journals that contain a high proportion of reviews tend to have high IFs. In contrast, clinical case reports tend to be cited sparsely.

Another problem is that journals are like zip codes. Suppose I live in 90210 (Beverly Hills, California, a wealthy suburb of Los Angeles). Am I wealthy? Wealth is not evenly distributed but follows a power law. I may be poor even though I live in a rich town. In fact, I may have moved to Beverly Hills just because I am hoping that my relatives would think I made it big. It is not unusual for businesses to rent post office boxes in prime locations to make customers think that they are well established and prosperous. Just like wealth, the citations that a journal garners follow a power law, and most of the citations come from a small proportion of its articles. In fact, most of its articles may be poorly cited or not cited at all. It is widely assumed that the IF is a good proxy for a journal's quality, both the quality of its peer review and the quality of its articles: Hogwash! That is simply not true. Each article needs to be assessed for its own merits, regardless of where it appears [6].

Still, authors are inclined to publish in journals that have high IFs so that their articles will bask in the journal's reputation. This is not silly, when you recall Ted Jones' admonition that promotion committees often judge a scientist according to the IFs of the journals that he or she has published in. I do find it surprising that the journal IF is used to rate individual articles, without considering how many citations that article actually got on its own! Hard to accept that hard-nosed scientists would let their lives be ruled by such spurious proxy measures.

Another metric based on citations, called the **H-Index**, is used to rate individual scientists by calculating the maximum number n such that the author has published n articles that have been cited at least n times. If I have published 10 papers that have been cited at least 10 times, my H-Index is 10. This takes into account productivity as well as citations, and gives some insight, though of course, the H-Index favors older, well-established researchers working in hot fields that can publish a lot of studies quickly.

An alternative article-based metric, the Relative Citation Ratio (RCR), has recently been proposed [7] that obviates many of the shortcomings of the IF, by calculating the number of citations per year made to an individual article, and comparing this to the average number of citations per year made to its co-citation network (Fig. 16.3). (To be more precise, they

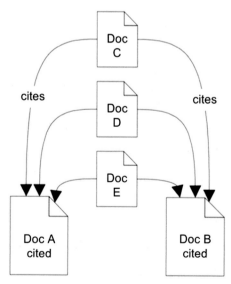

FIGURE 16.3   **Three articles C, D, and E cite A and B, so that A and B are co-cited.** *Reprinted from https://en. wikipedia.org/wiki/Co-citation.*

D. MAKE YOUR DATA GO FARTHER

compared it to the average journal IFs of the articles in the co-citation network, to have a "smoothing" effect in cases where the number of articles in the co-citation network is small.) The idea is that the RCR ranks citations to individual articles relative to a better reference baseline set that reflects its field and peer group.

Other article-based metrics (referred to collectively as **Altmetrics**) include counting how many times the article is downloaded from the journal or from repositories; how many times the article is referred to in blogs and tweets; how many users save or share the article in Mendeley or other academic websites; and how many times news stories about the article have appeared. Each of these metrics captures a different type of impact or influence.

NEW TRENDS IN PEER REVIEW

Peer review has come under attack for being slow, inefficient, potentially biased, and not being much better than random in rating the quality of submissions. As an editor for several journals, my philosophy is that peer review should not act as a gatekeeper for publication (unless the work is severely flawed). Nevertheless reviewers have a valuable role to play by giving suggestions that help the authors to improve their article until it is as complete and polished as possible.

The scientific community is currently trying a variety of experiments in peer review, for example:

1. Certain journals have an open peer-review policy in which the names of the reviewers are made public. Moreover, *Biology Direct*, and later *EMBO Journal,* and others, have begun to publish the full set of reviews and author replies, either at the end of the published article or in a file linked to the published article.
2. Other journals publish first and carry out peer review later. Besides *F1000Research* and *PeerJ*, mentioned above, the Wellcome Trust has launched a journal, *Wellcome Open*, in which any Wellcome-funded researcher can publish an article immediately prior to peer review. *Philica* goes a step further, since any academic researcher can publish literally any kind of article or observation, which is then open to be reviewed and/or commented on afterward.

There are also online venues being set up to allow readers to comment on published articles or even to carry out detailed postpublication peer review. Individual journals may allow reader comments to be linked to published articles, but these are not very popular and the comments are not indexed and hence are not findable by search engines. PubMed Commons is a public forum in which qualified readers (anyone who has authored an article indexed in PubMed) can post comments on any other PubMed article, and the comments are linked back to the article in PubMed. PubPeer allows anonymous comments, which has pros (people can critique an article without fear of reprisal) and cons (anyone is susceptible to unfair attack). Critiques posted on PubPeer have identified flaws in papers leading to retractions, but I think anonymity is at odds with the civility and transparency (Fig. 16.4) that the open science movement generally strives to achieve.

FIGURE 16.4　**Transparency is a good thing, in science at least.**

THE SCIENTIFIC ARTICLE AS A DATA OBJECT

In parallel to the changes occurring in the policies of journals, there are also many proposals for transforming the scientific article itself. Once an article is no longer viewed as a physical object on paper, but is digital (that is, a data object), then a published article is no longer static but can incorporate a series of updated new versions and even incorporate additional experiments. The article can accommodate hyperlinks and allow readers to mark up the text and share the annotations (e.g., the Hypothes.is project, https://hypothes.is/ [8]). Most importantly, the text can be deconstructed in terms of its underlying concepts, relationships, events, and assertions [9,10]. These innovations are transformational in at least three senses:

First, they allow scientists to mine text within and across the scientific literature, using the power of data science. Relevant articles on a given topic (or using the same reagents or models) can be retrieved much more readily. And knowledge can be assembled from assertions that reside in different articles or different disciplines, to create new promising hypotheses [11].

Second, the purpose of the scientific article is changing fundamentally. Traditionally, writing an article has been regarded as writing a letter to the scientific community, and a good article is one that tells a good story and has a strong narrative. This shifts when **data and evidence become the primary features of the article, instead of the author's interpretation**. Articles are no longer intended to be read only by humans—but also contain detailed representations to be read by computers as well (Fig. 16.5).

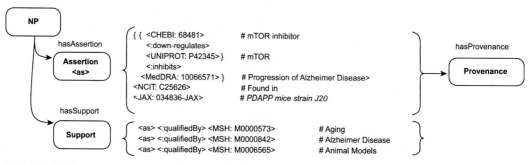

FIGURE 16.5 **Representation of statements and evidence in a nanopublication format.** The diagram shows an example of a so-called nanopublication (representing a self-contained assertion), which attempts to express the assertion that "inhibition of mTOR by rapamycin can slow or block AD progression in a transgenic mouse model of the disease." *Reprinted from Clark T, Ciccarese PN, Goble CA. Micropublications: a semantic model for claims, evidence, arguments and annotations in biomedical communications. J Biomed Semantics July 4, 2014;5:28. http://dx.doi.org/10.1186/2041-1480-5-28 with permission.*

Third, in my opinion, the whole notion of scientific progress is becoming altered. Scientific articles build on previous work and produce new findings and ideas; this process used to be compatible with the idea that science marches steadily toward certainty and truth. Ironically, the more that an article becomes deconstructed with the goal of laying its meaning bare and transparent, the more we appreciate that an article actually holds multiple possible meanings. The possible meanings evolve and change over time as the article receives multiple readings and multiple layers of commentary. A finding made in 2017 may be interpreted quite differently in 2027, in the light of other subsequent findings.

In short, scientific articles, despite retaining their rigor, validity, robustness, and reproducibility, are gradually being stripped of any pretense to **certainty** or **finality of interpretation**. A similar point was made by Jonathan Rosen when comparing Web pages on the Internet to entries in the Talmud, a Jewish religious book:

> I have often thought, contemplating a page of Talmud, that it bears a certain uncanny resemblance to a home page on the Internet, where nothing is whole in itself but where icons and text-boxes are doorways through which visitors pass into an infinity of cross-referenced texts and conversations. Consider a page of Talmud. There are a few lines of Mishnah, the conversation the Rabbis conducted (for some five hundred years before writing it down) about a broad range of legalistic questions stemming from the Bible but ranging into a host of other matters as well. Underneath those few lines begins the Gemarah, the conversation *later* Rabbis had about the conversation *earlier* Rabbis had in the Mishnah [12].

Perhaps an even better comparison is between the information-extracted, ontology-represented, hyperlinked scientific article and the text of *Finnegans Wake* (Box 16.1).

WHERE SHOULD I PUBLISH MY PAPER?

The marketplace of scholarly journals has never been bigger or more diverse. Literally any article can be published somewhere, and as long as the article is published in an open access format in a journal that is peer reviewed and indexed by the major services, it will be seen. However, people publish not simply to distribute the content of an article but to send a

BOX 16.1

THE FIRST SENTENCE OF THE NOVEL *FINNEGANS WAKE* [13], WITH WORDS LINKED TO COMMENTARY BELOW WHICH POINTS OUT THEIR MULTIPLE AMBIGUITIES AND LEVELS OF MEANING [14]

riverrun, past Eve and Adam's, from swerve of shore to bend of bay, brings us by a commodius vicus of recirculation back to Howth Castle and Environs [13].

riverrun - the course which a river shapes and follows through the landscape + The Letter: Reverend (letter start) + (Egyptian hieroglyphic) = 'rn' or 'ren' - name + Samuel Taylor Coleridge, Kubla Khan: "In Xanadu did Kubla Khan / A stately pleasure-dome decree: / Where Alph, the sacred river, ran / Through caverns measureless to man / Down to a sunless sea." (poem was composed one night after Coleridge experienced an opium influenced dream. Upon waking, he set about writing lines of poetry that came to him from the dream until he was interrupted. The poem could not be completed according to its original 200–300 line plan as the interruption caused him to forget the lines: "though he still retained some vague and dim recollection of the general purport of the vision, yet, with the exception of some eight or ten scattered lines and images, all the rest had passed away like the images on the surface of a stream into which a stone had been cast, but, alas! without the after restoration of the latter").

'Church of the Immaculate Conception', also known as Adam and Eve's, is located on Merchants Quay, Dublin (Franciscans secretly said Mass in the Adam and Eve Tavern, where the popular name of the present church comes from) + "Old as they were, her aunts also did their share. Julia, though she was quite grey, was still the leading soprano in Adam and Eve's, and Kate, being too feeble to go about much, gave music lessons to beginners on the old square piano in the back room." (The Dead); Miss Kate and Miss Julia, based on Joyce's grand aunts, the Misses Flynn who, as their great-nephew put it, "trilled and warbled in a Dublin church up to the age of seventy." This

was the ancient Franciscan church on the south quays popularly known as Adam and Eve's (from Biography by Peter Costello).

swerve - an abrupt change of direction, an erratic deflection from an intended course

bend - curve

bay - a body of water partially enclosed by land but with a wide mouth, affording access to the sea + Dublin Bay.

commodious - roomy and comfortable + Commodus - Roman Emperor from 180 to 192. The son of Marcus Aurelius, he is often considered to have been one of the worst Roman Emperors, and his reign brought to a close the era of the 'five good emperors'. He had a twin brother, Antoninus, who died when he was about four years old, and a sister Lucilla who was implicated in plots to overthrow him.

vicus (l) - village, hamlet; row of houses, quarter of a city + Giambattista Vico + vicious circle - situation in which a cause produces a result that itself produces the original cause → "The world of objects and solidity is a way of making our passage on earth convenient. It is only a description that was created to help us. Each of us, or rather our reason, forgets that the description is only a description, and thus we entrap the totality of ourselves in a vicious circle from which we rarely emerge in our lifetime." (Carlos Castaneda: Tales of Power)

recirculation - a renewed or fresh circulation

Howth - promontory and peninsula on the northern side of Dublin bay

environs - surroundings, outskirts + FDV (First Draft Version): brings us to Howth Castle & Environs! [14]

message about themselves! So as a practical matter, the proper choice of venue remains as important as ever.

One would think that the choice would be clear: publishing in a high-impact journal is most desirable, then first tier, second tier, and down the line. But before sending that article to *Science*, ask yourself: Is your finding hot? Do you have a track record of publications in this area already? Can you find a well-known scientist (preferably a member of the Editorial Board) to champion the work and help get it published?

In my opinion, more important than landing in a high-impact journal is to find a journal that matches your article and your target audience. Even in this age of online search engines, each scientist tends to have a short list of journals that they follow closely, and if you want to have your message appreciated by the peers in your field (or by the members of your promotion committee), you should place your article in one of those journals. Journals often host special issues on specific topics and announce a call for papers; these are good ways to gain the attention of workers in that specific area.

Tip: Publishing a great paper in a great journal can still be a bad move if it does not reach your target audience. *Journal of Experimental Medicine* is a first-tier journal but not the best place to build up "street cred" among molecular biologists [15].

Are you concerned about patents or being scooped? It is common for established investigators to publish short preliminary reports in some third-tier journal that will publish it rapidly, to establish priority, and then to submit a big definitive paper later to a first-tier journal. However, this might not be a good strategy for someone relatively unknown, and it may be worth fighting to get it into the first-tier journal in the first place.

IS THERE AN IDEAL PUBLISHING PORTFOLIO?

In discussing where to publish an article, we should also consider how any single article sits in the life cycle of a scientific career. An oversimplified, linear career model assumes that a young scientist (1) chooses a discipline and a mentor, (2) trains until he or she is ready to establish his/her own lab, (3) gets an independent tenure track academic position in the same discipline, (4) finds a niche or research program, (5) builds a portfolio of publications, and (6) rises through the ranks. Science metrics such as RCR and H-Index implicitly reward scientists who follow this mold, but tend to underestimate the impact of scientists who move from academia to industry or government, or move from discipline to discipline. Such scientists may have a wide circle of colleagues and collaborators, yet not have a stable set of peers over the long term who are publishing related works and who are likely to cite them consistently.

The most popular science metrics all assign some average impact value to a scientist, assuming that ideally each article that they write ought to have as high an impact as possible (measured in citations). What is wrong with that? If John Lennon had been a scientist, and if his metrics measured record sales instead of citations, then he would get big points for *Imagine*, but would be dinged for *How do you Sleep?* and many others. That is, he would be penalized for every song he wrote that did not sell well. Yet intuitively, we reject that approach—we realize that, instead, he should be *further* rewarded for his sheer creativity, diversity, productivity, and risk taking, regardless of the sales figures. We understand that the artistic process **involves letting 1000 flowers bloom**.

In fact, many leading scientists behave in the same manner. They let 1000 flowers bloom, write papers of all kinds, and publish in all kinds of venues. I think of the neuroscientist Vilayanur S. Ramachandran, who to date has published 22 articles in a high-impact journal (*Nature*) and 21 in *Medical Hypotheses*, a fourth-tier journal. Or Mark P. Mattson, who published an experiment done with his teenage son in which they soaked nails in a solution containing beta-amyloid peptide to see if it would accelerate rusting [16]. Or, my old colleague, the late Don Swanson, who trained as a theoretical physicist, was dean of the Graduate Library School at University of Chicago, and wrote a series of articles on psychoanalysis, including a hilarious spoof [17].

Traditional metrics, and conventional wisdom, would penalize these investigators. Yet they survived, and thrived. How? They each established a niche for themselves and published extensively in the leading scholarly journals of their own disciplines. Their research stood the test of time in terms of validity, robustness, and reproducibility. They introduced new concepts and approaches and opened up new fields of inquiry. I have written this book to give you, the reader, the tools that will allow you to follow in their footsteps.

References

[1] Walker R, Rocha da Silva P. Emerging trends in peer review — a survey. Front Neurosci May 27, 2015;9:169. http://dx.doi.org/10.3389/fnins.2015.00169.

[2] Campanario JM. Scientometrics 2009;81:549. http://dx.doi.org/10.1007/s11192-008-2141-5.

[3] McCulloch WS, Pitts W. Bull Math Biophys 1943;5:115. http://dx.doi.org/10.1007/BF02478259.

[4] http://www.budapestopenaccessinitiative.org/.

[5] Garfield E. The history and meaning of the journal impact factor. JAMA January 4, 2006;295(1):90—3.

[6] Barbui C, Cipriani A, Malvini L, Tansella M. Validity of the impact factor of journals as a measure of randomized controlled trial quality. J Clin Psychiatry January 2006;67(1):37—40.

[7] Hutchins BI, Yuan X, Anderson JM, Santangelo GM. Relative Citation Ratio (RCR): a new metric that uses citation rates to measure influence at the article level. PLoS Biol September 6, 2016;14(9):e1002541.

[8] Martone M, Murray-Rust P, Molloy J, Arrow T, MacGillivray M, Kittel C, Kasberger S, Steel G, Oppenheim C, Ranganathan A, Tennant J, Udell J. ContentMine/hypothesis proposal. Res Ideas Outcomes 2016;2:e8424. https://doi.org/10.3897/rio.2.e8424.

[9] Przybyła P, Shardlow M, Aubin S, Bossy R, Eckart de Castilho R, Piperidis S, McNaught J, Ananiadou S. Text mining resources for the life sciences. Database (Oxford) November 25, 2016;2016. pii: baw145.

[10] Clark T, Ciccarese PN, Goble CA. Micropublications: a semantic model for claims, evidence, arguments and annotations in biomedical communications. J Biomed Semantics July 4, 2014;5:28. http://dx.doi.org/10.1186/2041-1480-5-28.

[11] Swanson DR, Smalheiser NR. An interactive system for finding complementary literature: a stimulus to scientific discovery. Artif Intell April 30, 1997;91(2):183—203.

[12] Rosen J. The Talmud and the internet. Am Scholar April 1, 1998;67(2):47—54.

[13] Joyce J. Finnegans Wake. New York and London: Penguin; 1939. 1999.

[14] http://www.finwake.com/1024chapter1/1024finn1.htm.

[15] Avery OT, MacLeod CM, McCarty M. Studies on the chemical nature of the substance inducing transformation of pneumococcal types: induction of transformation by a desoxyribonucleic acid fraction isolated from pneumococcus type III. J Exp Med February 1, 1944;79(2):137.

[16] Mattson MP, Mattson EP. Amyloid peptide enhances nail rusting: novel insight into mechanisms of aging and Alzheimer's disease. Ageing Res Rev June 2002;1(3):327—30.

[17] Swanson DR. New horizons in psychoanalysis: treatment of necrosistic personality disorders. Perspect Biol Med 1986;29(4):493—8.

Postpublication peer review is not always this kind.

Postscript: Beyond Data Literacy

My intent in writing this book has been to give the reader a "feel" for designing experiments and handling data and to pass on tips that I have learned over more than 30 years of doing bench science.

I consider myself data literate, but not data competent. Almost all of my studies are done in collaboration with data scientists of one kind or another. At present, there are relatively few individuals who have advanced training both in a domain (such as biology or social sciences) and in data science (including computer science, machine learning, informatics, and databases). Yet such individuals are in great demand in the job market, and I predict that in the near future, data competence will no longer be an exotic skill but a basic job requirement.

Becoming data literate is just the beginning. To carry out research studies in the laboratory, clinic, or the field, you will need to take more specialized courses in statistics and experimental design. And ultimately, there is a big digital divide that you will need to cross: I refer to the growing chasm between those who can program computers versus those who cannot. You can do a lot by entering data into spreadsheets, software, and Web forms, but to do any serious data science, or to handle big data of any kind, it is necessary to set up your own databases and write your own code.

It is time to take the concepts you have learned off the printed page and use them in real life!

Index

Printed in the United States
By Bookmasters